Signal Processing for Neuroscientists, A Companion Volume

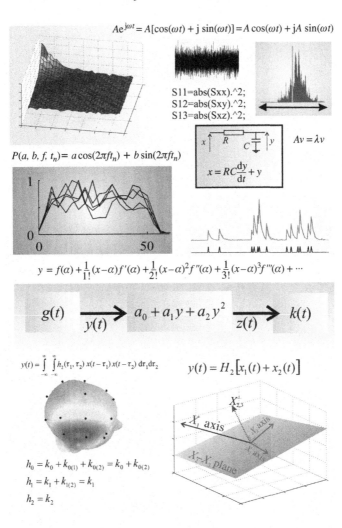

Signal Processing for Neuroscientists, A Companion Volume

Advanced Topics, Nonlinear Techniques and Multi-Channel Analysis

Wim van Drongelen

ELSEVIER

AMSTERDAM • BOSTON • HEIDELBERG • LONDON • NEW YORK • OXFORD
PARIS • SAN DIEGO • SAN FRANCISCO • SINGAPORE • SYDNEY • TOKYO

Elsevier
32 Jamestown Road London NW1 7BY
30 Corporate Drive, Suite 400, Burlington, MA 01803, USA

First edition 2010

Notices
Knowledge and best practice in this field are constantly changing. As new research and experience
broaden our understanding, changes in research methods, professional practices, or medical treatment
may become necessary.

Practitioners and researchers must always rely on their own experience and knowledge in evaluating and
using any information, methods, compounds, or experiments described herein. In using such information
or methods they should be mindful of their own safety and the safety of others, including parties for
whom they have a professional responsibility.

To the fullest extent of the law, neither the Publisher nor the authors, contributors, or editors, assume
any liability for any injury and/or damage to persons or property as a matter of products liability,
negligence or otherwise, or from any use or operation of any methods, products, instructions, or ideas
contained in the material herein.

British Library Cataloguing-in-Publication Data
A catalogue record for this book is available from the British Library

Library of Congress Cataloging-in-Publication Data
A catalog record for this book is available from the Library of Congress

ISBN: 978-0-323-16514-3

For information on all Elsevier publications
visit our website at elsevierdirect.com

This book has been manufactured using Print On Demand technology. Each copy is produced to order
and is limited to black ink. The online version of this book will show color figures where appropriate.

Working together to grow
libraries in developing countries

www.elsevier.com | www.bookaid.org | www.sabre.org

ELSEVIER   BOOK AID International   Sabre Foundation

Contents

Preface   **vii**

1   Lomb's Algorithm and the Hilbert Transform   **1**
    1.1   Introduction   1
    1.2   Unevenly Sampled Data   1
    1.3   The Hilbert Transform   8
    Appendix 1.1   17
    Appendix 1.2   18
    Appendix 1.3   19

2   Modeling   **21**
    2.1   Introduction   21
    2.2   Different Types of Models   21
    2.3   Examples of Parametric and Nonparametric Models   23
    2.4   Polynomials   26
    2.5   Nonlinear Systems with Memory   32
    Appendix 2.1   36

3   Volterra Series   **39**
    3.1   Introduction   39
    3.2   Volterra Series   42
    3.3   A Second-Order Volterra System   45
    3.4   General Second-Order System   51
    3.5   System Tests for Internal Structure   53
    3.6   Sinusoidal Signals   57

4   Wiener Series   **61**
    4.1   Introduction   61
    4.2   Wiener Kernels   62
    4.3   Determination of the Zero-, First-, and Second-Order Wiener Kernels   69
    4.4   Implementation of the Cross-Correlation Method   73
    4.5   Relation between Wiener and Volterra Kernels   77
    4.6   Analyzing Spiking Neurons Stimulated with Noise   79
    4.7   Nonwhite Gaussian Input   84

4.8 Summary   86
Appendix 4.1   87
Appendix 4.2   89

5   Poisson–Wiener Series   **91**
5.1 Introduction   91
5.2 Systems with Impulse Train Input   91
5.3 Determination of the Zero-, First-, and Second-Order
     Poisson–Wiener Kernels   103
5.4 Implementation of the Cross-Correlation Method   109
5.5 Spiking Output   111
5.6 Summary   112
Appendix 5.1   113
Appendix 5.2   117

6   Decomposition of Multichannel Data   **119**
6.1 Introduction   119
6.2 Mixing and Unmixing of Signals   120
6.3 Principal Component Analysis   123
6.4 Independent Component Analysis   134
Appendix 6.1   155

7   Causality   **159**
7.1 Introduction   159
7.2 Granger Causality   160
7.3 Directed Transfer Function   160
7.4 Combination of Multichannel Methods   175

References   **177**

Preface

This text is based on a course I teach at the University of Chicago for students in Computational Neuroscience. It is a continuation of the previously published text *Signal Processing for Neuroscientists: An Introduction to the Analysis of Physiological Signals* and includes some of the more advanced topics of linear and nonlinear systems analysis and multichannel analysis. In the following, it is assumed that the reader is familiar with the basic concepts that are covered in the introductory text and, to help the student, multiple references to the basics are included.

The popularity of signal processing in neuroscience is increasing, and with the current availability and development of computer hardware and software it may be anticipated that the current growth will continue. Because electrode fabrication has improved and measurement equipment is getting less expensive, electrophysiological measurements with large numbers of channels are now very common. In addition, neuroscience has entered the age of light, and fluorescence measurements are fully integrated into the researcher's toolkit. Because each image in a movie contains multiple pixels, these measurements are multichannel by nature. Furthermore, the availability of both generic and specialized software packages for data analysis has altered the neuroscientist's attitude toward some of the more complex analysis techniques. Interestingly, the increased accessibility of hardware and software may lead to a rediscovery of analysis procedures that were initially described decades ago. At the time when these procedures were developed, only few researchers had access to the required instrumentation, but now most scientists can access both the necessary equipment and modern computer hardware and software to perform complex experiments and analyses.

The considerations given above have provided a strong motivation for the development of this text, where we discuss several advanced techniques, rediscover methods to describe nonlinear systems, and examine the analysis of multichannel recordings. The first chapter describes two very specialized algorithms: Lomb's algorithm to analyze unevenly sampled data sets and the Hilbert transform to detect instantaneous phase and amplitude of a signal. The remainder of the text can be divided into two main components: (I) modeling systems (Chapter 2) and the analysis of nonlinear systems with the Volterra and Wiener series (Chapters 3−5) and (II) the analysis of multichannel measurements using a statistical approach (Chapter 6) and examination of causal relationships (Chapter 7). Throughout this text, we adopt an informal approach to the development of algorithms and we include practical examples implemented in MATLAB. (All the MATLAB scripts used in this text can be obtained via http://www.elsevierdirect.com/companions/9780123849151)

It is a pleasure to acknowledge those who have assisted (directly and indirectly) in the preparation of this text: Drs. V.L. Towle, P.S. Ulinski, D. Margoliash, H.C. Lee, M.H. Kohrman, P. Adret, and N. Hatsopoulos. I also thank the teaching assistants for their help in the course and in the development of the material in this text: thanks, Matt Green, Peter Kruskal, Chris Rishel, and Jared Ostmeyer. There is a strong coupling between my teaching efforts and research interests. Therefore, I am indebted to the Dr. Ralph and Marian Falk Medical Research Trust for supporting my research and to the graduate and undergraduate students in my laboratory: Jen Dwyer, Marc Benayoun, Amber Martell, Mukta Vaidya, and Valeriya Talovikova. They provided useful feedback, tested some of the algorithms, and collected several example data sets. Special thanks to the group of students in the 2010 winter class who helped me with reviewing this material: Matt Best, Kevin Brown, Jonathan Jui, Matt Kearney, Lane McIntosh, Jillian McKee, Leo Olmedo, Alex Rajan, Alex Sadovsky, Honi Sanders, Valeriya Talovikova, Kelsey Tupper, and Richard Williams. Their multiple suggestions and critical review helped to significantly improve the text and some of the figures. At Elsevier I want to thank Lisa Tickner, Clare Caruana, Lisa Jones, Mani Prabakaran, and Johannes Menzel for their help and advice. Last but not least, thanks to my wife Ingrid for everything and supporting the multiple vacation days used for writing.

1 Lomb's Algorithm and the Hilbert Transform

1.1 Introduction

This first chapter describes two of the more advanced techniques in signal processing: Lomb's algorithm and the Hilbert transform. Throughout this chapter (and the remainder of this text) we assume that you have a basic understanding of signal processing procedures; for those needing to refresh these skills, we include multiple references to van Drongelen (2007).

In the 1970s, the astrophysicist Lomb developed an algorithm for spectral analysis to deal with signals consisting of unevenly sampled data. You might comment that in astrophysics considering uneven sampling is highly relevant (you cannot observe the stars on a cloudy day), but in neuroscience data are always evenly sampled. Although this is true, one can consider the action potential (or its extracellular recorded equivalent, the spike) or neuronal burst as events that represent or sample an underlying continuous process. Since these events occur unevenly, the sampling of the underlying process is also uneven. In this context we will explore how to obtain spectral information from unevenly distributed events.

The second part of this chapter introduces the Hilbert transform that allows one to compute the instantaneous phase and amplitude of a signal. The fact that one can determine these two metrics in an instantaneous fashion is unique because usually this type of parameter can only be associated with an interval of the signal. For example, in spectral analysis the spectrum is computed for an epoch and the spectral resolution is determined by epoch length. Being able to determine parameters such as the phase instantaneously is especially useful if one wants to determine relationships between multiple signals generated within a neuronal network.

1.2 Unevenly Sampled Data

In most measurements we have evenly sampled data—for instance, the interval Δt between the sample points of the time series is constant, pixels in a picture have uniform interdistance, and so forth. Usually this is the case, but there are instances when uneven sampling cannot be avoided. Spike trains (chapter 14, van Drongelen,

Signal Processing for Neuroscientists, A Companion Volume. DOI: 10.1016/B978-0-12-384915-1.00001-2

2007) or time series representing heart rate (van Drongelen et al., 2009) are two such examples; in these cases one may consider the spike or the heartbeat to represent events that sample an underlying process that is invisible to the experimenter (Fig. 1.1A).

The heart rate signal is usually determined by measuring the intervals between peaks in the QRS complexes. The inverse value of the interval between pairs of subsequent QRS complexes can be considered a measure of the instantaneous rate (Fig. 1.1B). This rate value can be positioned in a time series at the instant of either the first or second QRS complex of the pair and, because the heartbeats do occur at slightly irregular intervals, the time series is sampled unevenly. This example for the heartbeat could be repeated, in a similar fashion, for determining the firing rate associated with a spike train.

When a signal is unevenly sampled, many algorithms that are based on a fixed sample interval (such as the direct Fourier transform [DFT] or fast Fourier transform [FFT]) cannot be applied. In principle there are several solutions to this problem:

(1) An evenly sampled time series can be constructed from the unevenly sampled one by using interpolation. In this approach the original signal is resampled at evenly spaced intervals. The interpolation technique (e.g., linear, cubic, spline) may vary with the application. In MATLAB resampling may be accomplished with the `interp1` command or any of the other related functions. After resampling the time series one can use standard Fourier analysis methods. The disadvantage is that the interpolation algorithm may introduce frequency components that are not related to the underlying process.

(2) The measurements can be represented as the number of events in a binned trace; now our time series is a sequence of numbers, with one number for each bin. Since the bins are equally spaced, the standard DFT/FFT can be applied. In case of low-frequency activity, the bins must be relatively wide to avoid an overrepresentation of empty bins.

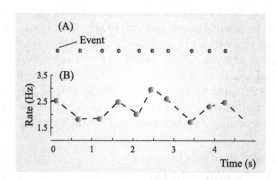

Figure 1.1 The QRS complexes in the ECG or extracellularly recorded spike trains can be considered as a series of events such as shown in (A). The rate of events can be depicted as the inverse of the interval between the events (B); here the inverse of the interval between each pair of events is plotted at the instant of the second event of the pair. The signal in (B) is unevenly sampled because the rate measure is available only at the occurrence of the events; the dashed line is a linear interpolation between these measures. *(Color in electronic version.)*

The disadvantage of this is that wide bins are associated with a low sample rate and thus a low Nyquist frequency, which limits the bandwidth of the spectral analysis.

(3) The most elegant solution is to use Lomb's algorithm for estimating the spectrum. This algorithm is specially designed to deal with unevenly sampled time series directly without the assumptions demanded by interpolation and resampling techniques (Lomb, 1976; Press et al., 1992; Scargle, 1982; van Drongelen et al., 2009). The background and application of this algorithm will be further described in Sections 1.2.1 and 1.2.2.

1.2.1 Lomb's Algorithm

The idea of Lomb's algorithm is similar to the development of the Fourier series, namely, to represent a signal by a sum of sinusoidal waves (see chapter 5 in van Drongelen, 2007). Lomb's procedure is to fit a demeaned time series x that **may be sampled unevenly** to a weighted pair of cosine and sine waves, where the cosine is weighted by coefficient a and the sine by coefficient b. The fitting procedure is performed over N samples of $x(n)$ obtained at times t_n and repeated for each frequency f.

$$\boxed{P(a,b,f,t_n) = a \cos(2\pi f t_n) + b \sin(2\pi f t_n)} \tag{1.1}$$

Coefficients a and b are unknown and must be obtained from the fitting procedure. For example, we can fit P to signal x by minimizing the squared difference between them over all samples: that is, minimize $\varepsilon^2 = \sum_{n=0}^{N-1} [P - X(n)]^2$. We repeat this minimization for each frequency f. To accomplish this, we follow the same procedure for developing the Fourier series (chapter 5 in van Drongelen, 2007) and set the partial derivative for each coefficient to zero to find the minimum of the error, that is:

$$\partial \varepsilon^2 / \partial a = 0 \tag{1.2a}$$

and

$$\partial \varepsilon^2 / \partial b = 0 \tag{1.2b}$$

For convenience, in the following we use a shorthand notation in addition to the full notation. In the shorthand notation: $C = \cos(2\pi f t_n)$, $S = \sin(2\pi f t_n)$, and $X = x(n)$.

For the condition in Equation (1.2a) we get:

$$\partial \varepsilon^2 / \partial a = \sum 2[P - x(n)] \frac{\partial [P - x(n)]}{\partial a} = \sum 2(aC + bS - X)C$$

$$= \sum_{n=0}^{N-1} 2 \left[\underbrace{a \cos(2\pi f t_n) + b \sin(2\pi f t_n)}_{P} - x(n) \right] \underbrace{\cos(2\pi f t_n)}_{\partial [P - x(n)]/\partial a} = 0$$

This and a similar expression obtained from the condition in Equation (1.2b) results in the following two equations:

$$\sum XC = a \sum C^2 + b \sum CS$$

$$\sum_{n=0}^{N-1} X(n)\cos(2\pi f t_n) = a \sum_{n=0}^{N-1} \cos^2(2\pi f t_n) + b \sum_{n=0}^{N-1} \cos(2\pi f t_n)\sin(2\pi f t_n) \qquad (1.3a)$$

and

$$\sum XS = a \sum CS + b \sum S^2$$

$$\sum_{n=0}^{N-1} X(n)\sin(2\pi f t_n) = a \sum_{n=0}^{N-1} \cos(2\pi f t_n)\sin(2\pi f t_n) + b \sum_{n=0}^{N-1} \sin^2(2\pi f t_n) \qquad (1.3b)$$

Thus far the procedure is similar to the standard Fourier analysis described in chapter 5 in van Drongelen (2007). The special feature in Lomb's algorithm is that for each frequency f, the sample times t_n are now shifted by an amount τ (Fig. 1.2). Thus, in Equations (1.3a) and (1.3b), t_n becomes $t_n - \tau$. The critical step is that for each frequency f, we select an optimal time shift τ so that the cosine–sine cross-terms ($\sum CS$) disappear, that is:

$$\sum CS = \sum_{n=0}^{N-1} \cos(2\pi f(t_n - \tau))\sin(2\pi f(t_n - \tau)) = 0 \qquad (1.4)$$

Using the trigonometric identity $\cos(A)\sin(B) = \frac{1}{2}[\sin(A - B) - \sin(A + B)]$, this can be simplified into:

$$\frac{1}{2}\left[\sum_{n=0}^{N-1} \underbrace{\sin(0)}_{0} - \sin(4\pi f(t_n - \tau)) \right] = 0 \rightarrow \sum_{n=0}^{N-1} \sin(4\pi f(t_n - \tau)) = 0$$

To separate the expressions for t_n and τ, we use the trigonometric relationship $\sin(A - B) = \sin(A)\cos(B) - \cos(A)\sin(B)$ to get the following expression:

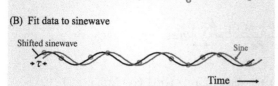

(A) Unevenly sampled signal

(B) Fit data to sinewave

Shifted sinewave

τ

Sine

Time ⟶

Figure 1.2 The Lomb algorithm fits sinusoidal signals to time series that may be unevenly sampled, as in the example in (A). The fit procedure (B) is optimized by shifting the sinusoidal signals by an amount τ. *(Color in electronic version.)*

$$\sum_{n=0}^{N-1}\sin(4\pi f t_n)\cos(4\pi f\tau) - \sum_{n=0}^{N-1}\cos(4\pi f t_n)\sin(4\pi f\tau)$$

$$= \cos(4\pi f\tau)\sum_{n=0}^{N-1}\sin(4\pi f t_n) - \sin(4\pi f\tau)\sum_{n=0}^{N-1}\cos(4\pi f t_n) = 0$$

This can be further simplified into:

$$\sin(4\pi f\tau)/\cos(4\pi f\tau) = \tan(4\pi f\tau) = \sum_{n=0}^{N-1}\sin(4\pi f t_n)\bigg/\sum_{n=0}^{N-1}\cos(4\pi f t_n)$$

Hence, condition (1.4) is satisfied if:

$$\tau = \tan^{-1}\left[\sum_{n=0}^{N-1}\sin(4\pi f t_n)\bigg/\sum_{n=0}^{N-1}\cos(4\pi f t_n)\right]\bigg/4\pi f \qquad (1.5)$$

The value of variable τ as a function of frequency f can be found with Equation (1.5), and by applying the appropriate shift $t_n \to (t_n-\tau)$, the cross-terms in Equations (1.3a) and (1.3b) become zero. Now we can determine the a and b coefficients for each frequency from the simplified expressions obtained from Equations (1.3a) and (1.3b) without the cross-terms:

$$\sum XC = a\sum C^2$$

$$\sum_{n=0}^{N-1}X(n)\cos(2\pi f(t_n-\tau)) = a\sum_{n=0}^{N-1}\cos^2(2\pi f(t_n-\tau)) \qquad (1.6a)$$

$$\to a = \sum XC\bigg/\sum C^2 = \sum_{n=0}^{N-1}x(n)\cos(2\pi f(t_n-\tau))\bigg/\sum_{n=0}^{N-1}\cos^2(2\pi f(t_n-\tau))$$

and

$$\sum XS = b\sum S^2$$

$$\sum_{n=0}^{N-1}x(n)\sin(2\pi f(t_n-\tau)) = b\sum_{n=0}^{N-1}\sin^2(2\pi f(t_n-\tau)) \qquad (1.6b)$$

$$\to b = \sum XS\bigg/\sum S^2 = \sum_{n=0}^{N-1}x(n)\sin(2\pi f(t_n-\tau))\bigg/\sum_{n=0}^{N-1}\sin^2(2\pi f(t_n-\tau))$$

Now we can compute the sum of $P^2(a, b, f, t_n)$—that is, the sum of squares of the sinusoidal signal in Equation (1.1) for all t_n—in order to obtain an expression that is proportional with the power spectrum S of $x(n)$ as a function of f:

$$S(f, a, b) = \sum_{n=0}^{N-1} P^2(a, b, f, t_n) = \sum (aC + bS)^2 = \sum a^2 C^2 + b^2 S^2 + \overbrace{2abCS}^{\text{cross-terms}}$$

$$= \sum_{n=0}^{N-1} \left[a^2 \cos^2(2\pi f(t_n - \tau)) + b^2 \sin^2(2\pi f(t_n - \tau)) + \underbrace{\text{cross-terms}}_{0} \right]$$

(1.7)

Since we shift by τ, all cross-terms vanish and by substitution of the expressions for the a and b coefficients in Equation (1.7) we get:

$$S(f) = \frac{\left(\sum XC\right)^2}{\left(\sum C^2\right)^2} \sum C^2 + \frac{\left(\sum XS\right)^2}{\left(\sum S^2\right)^2} \sum S^2$$

$$= \frac{\left[\sum_{n=0}^{N-1} x(n)\cos(2\pi f(t_n - \tau))\right]^2}{\left[\sum_{n=0}^{N-1} \cos^2(2\pi f(t_n - \tau))\right]^2} \sum_{n=0}^{N-1} \cos^2(2\pi f(t_n - \tau))$$

$$+ \frac{\left[\sum_{n=0}^{N-1} x(n)\sin(2\pi f(t_n - \tau))\right]^2}{\left[\sum_{n=0}^{N-1} \sin^2(2\pi f(t_n - \tau))\right]^2} \sum_{n=0}^{N-1} \sin^2(2\pi f(t_n - \tau))$$

This can be further simplified into:

$$S(f) = \frac{\left(\sum XC\right)^2}{\sum C^2} + \frac{\left(\sum XS\right)^2}{\sum S^2}$$

$$= \frac{\left[\sum_{n=0}^{N-1} x(n)\cos(2\pi f(t_n - \tau))\right]^2}{\sum_{n=0}^{N-1} \cos^2(2\pi f(t_n - \tau))} + \frac{\left[\sum_{n=0}^{N-1} x(n)\sin(2\pi f(t_n - \tau))\right]^2}{\sum_{n=0}^{N-1} \sin^2(2\pi f(t_n - \tau))}$$

(1.8)

The expression for the power spectrum in Equation (1.8) is sometimes divided by 2 (to make it equal to the standard power spectrum based on the Fourier transform; see Appendix 1.1), or by $2\sigma^2$ (σ^2—variance of x) for the determination of the statistical significance of spectral peaks. (Some of the background for this normalization is described in Appendix 1.1; for more details, see Scargle, 1982.) By applying the normalization we finally get:

$$
S(f) = \frac{1}{2\sigma^2} \left\{ \frac{\left[\sum_{n=0}^{N-1} x(n)\cos(2\pi f(t_n - \tau)) \right]^2}{\sum_{n=0}^{N-1} \cos^2(2\pi f(t_n - \tau))} + \frac{\left[\sum_{n=0}^{N-1} x(n)\sin(2\pi f(t_n - \tau)) \right]^2}{\sum_{n=0}^{N-1} \sin^2(2\pi f(t_n - \tau))} \right\}
$$

(1.9)

From the above derivation, we can see that Lomb's procedure allows (but does not require) unevenly sampled data. Note that in Equations (1.7) and (1.8) we did not compute power as the square of the cosine and sine coefficients, a and b, as we would do in the standard Fourier transform; this is because in Lomb's approach the sinusoidal signals are not required to have a complete period within the epoch determined by the samples $x(n)$. Because we do not have this requirement, the frequency f is essentially a continuous variable and the spectral estimate we obtain by this approach is therefore not limited by frequency resolution (in the DFT/FFT, the frequency resolution is determined by the total epoch of the sampled data) and range (in the DFT/FFT, the maximum frequency is determined by the Nyquist frequency). However, to avoid misinterpretation, it is common practice to limit the bandwidth of the Lomb spectrum to less than or equal to half the average sample rate. Similarly, the commonly employed frequency resolution is the inverse of the signal's epoch.

1.2.2 A MATLAB Example

To test Lomb's algorithm we apply it to a signal that consists of two sinusoidal signals (50 and 130 Hz) plus a random noise component (this is the same example used in fig. 7.2A in van Drongelen, 2007). In this example (implemented in MATLAB script Pr1_1.m), we sample the signal with randomly distributed intervals (2000 points) and specify a frequency scale (f in the script) up to 500 Hz. Subsequently we use Equations (1.5), (1.8), and (1.9) to compute τ (tau in the script) and the unscaled and scaled versions of power spectrum $S(f)$ (Pxx in the script) of input $x(n)$ (x in the script). This script is available on http://www.elsevierdirect.com/companions/9780123849151.

The following script (Pr1_1.m) uses the Lomb algorithm to compute the spectrum from an unevenly sampled signal. The output of the script is a plot of the input (an unevenly sampled time domain) signal and its associated Lomb spectrum.

```
% Pr1_1.m
% Application of Lomb Spectrum
clear;

t=rand(2000,1); t=sort(t);      % An array of 2000 random sample intervals
f=[1:500];                      % The desired frequency scale
% frequencies same as pr7_1.m in van Drongelen (2007)
f1=50;
f2=130;
% data plus noise as in pr7_1.m in van Drongelen (2007)
x=sin(2*pi*f1*t)+sin(2*pi*f2*t);
x=x+randn(length(t),1);
var=(std(x))^2;                 % The signal's variance

% Main Loop
for i=1:length(f)
   h1=4*pi*f(i)*t;
   %Equation (1.5)
   tau=atan2(sum(sin(h1)), sum(cos(h1)))/(4*pi*f(i));
   h2=2*pi*f(i)*(t-tau);
   %Equation (1.8)
   Pxx(i)=(sum(x.*cos(h2)).^2)/sum(cos(h2).^2)+...
                         (sum(x.*sin(h2)).^2)/sum(sin(h2).^2);
end;
% Normalize; Equation (1.9)
Pxx=Pxx/(2*var);
% Plot the Results
figure;
subplot(2,1,1), plot(t,x,'.−')
title('Irregularly Sampled Signal (USE ZOOM TO INSPECT UNEVEN
SAMPLING)')
xlabel('Time (s)');ylabel('Amplitude')
subplot(2,1,2),plot(f,Pxx);
title('Lomb Spectrum')
xlabel('Frequency (Hz)');ylabel('Normalized Power')
```

1.3   The Hilbert Transform

One of the current frontiers in neuroscience is marked by our lack of understanding of neuronal network function. A first step in unraveling network activities is to record from multiple neurons and/or networks simultaneously. A question that often arises in this context is which signals lead or lag; the underlying thought here is that the signals that lead cause the signals that lag. Although this approach is not foolproof, since one can only make reasonable inferences about causality if all connections between and activities of the neuronal elements are established, it is a first step in analyzing network function. Multiple techniques to measure lead and lag can be used. The simplest ones are cross-correlation and coherence (for an overview of these techniques, see chapter 8 in van Drongelen, 2007). A rather direct method to examine lead and lag is to determine the phase of simultaneously recorded signals. If the phase difference between two signals is not too big, one considers signal 1 to lead signal 2 if the phase of signal 1 (ϕ_1) is less than the phase of signal 2 (ϕ_2): $\phi_1 < \phi_2$. Of course this procedure should be considered as a heuristic approach to describe a causal sequence between the components in the network activity since there is no guarantee that a phase difference reflects a causal relationship between neural element 1 (generating signal 1) and neural element 2 (generating signal 2). In this example one could easily imagine alternatives where neural elements 1 and 2 are both connected to a common source causing both signals, or where element 2 is connected to element 1 via a significant number of relays; in both alternatives the condition $\phi_1 < \phi_2$ might be satisfied without a direct causal relationship from element 1 to element 2. A frequently used technique to compute a signal's phase is the Hilbert transform, which will be described in the remainder of this chapter. An alternative approach to study causality in multichannel data is discussed in Chapter 7.

The Hilbert transform is a useful tool to determine the amplitude and instantaneous phase of a signal. We will first define the transform before demonstrating the underlying mathematics. An easy way of introducing the application of the Hilbert transform is by considering Euler's equation multiplied with a constant A:

$$A\mathrm{e}^{\mathrm{j}\omega t} = A[\cos(\omega t) + \mathrm{j}\sin(\omega t)] = A\cos(\omega t) + \mathrm{j}A\sin(\omega t) \tag{1.10}$$

In this example we consider the first term in Equation (1.10), $f(t) = A\cos(\omega t)$, as the **signal** under investigation. This signal is ideal to demonstrate the Hilbert transform application because in this example we can see that the amplitude of $f(t)$ is A, and its instantaneous phase ϕ is ωt. The terminology for the Hilbert transform is as follows: the imaginary component, the second term, in Equation (1.10) $\tilde{f}(t) = A\sin(\omega t)$ is defined as the **Hilbert transform** of $f(t)$ (we will discuss further details in Sections 1.3.1 and 1.3.2 below), and the sum of both the signal and its Hilbert transform multiplied by j generates a complex signal:

$$f_a(t) = A\,\mathrm{e}^{\mathrm{j}\omega t} = A\cos(\omega t) + \mathrm{j}\,A\sin(\omega t) = f(t) + \mathrm{j}\tilde{f}(t)$$

in which $f_a(t)$ is defined as the **analytic signal**.

To summarize, the real part of the analytic signal is the signal under investigation $f(t)$ and its imaginary component is the Hilbert transform $\tilde{f}(t)$ of the signal. The analysis procedure is summarized in Fig. 1.3. As can be seen in Fig. 1.3A and B, we can use the analytic signal $A\,\mathrm{e}^{\mathrm{j}\omega t}$ to determine amplitude A and instantaneous phase ωt of any point, such as the one indicated by *. The amplitude is:

$$A = \sqrt{\text{real component}^2 + \text{imaginary component}^2}$$

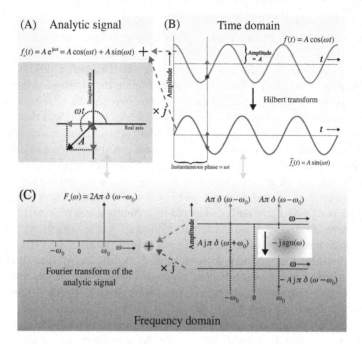

Figure 1.3 The signal amplitude A and instantaneous phase ωt of point * of the cosine function (B, $f(t)$, blue) can be determined with the so-called analytic signal (A). The analytic signal consists of a real part equal to the signal under investigation (the cosine) and an imaginary component (the sine). The imaginary component (red) is defined as the Hilbert transform $\tilde{f}(t)$ of the signal $f(t)$. The frequency domain equivalents of the cosine wave, the sine wave, the Hilbert transform procedure, and the analytic signal are shown in (C). See text for further explanation. *(Color in electronic version.)*

and the phase is:

$$\phi = \tan^{-1}\left[\frac{\text{imaginary component}}{\text{real component}}\right]$$

Again, in this example we did not need the analytic signal to determine phase and amplitude for our simple cosine signal, but our finding may be generalized to other signals where such a determination is not trivial.

1.3.1 The Hilbert Transform in the Frequency Domain

As can be seen in the earlier example (depicted in Fig. 1.3), the Hilbert transform can be considered as a phase shift operation on $f(t)$ to generate $\tilde{f}(t)$. In our example the signal $\cos(\omega t)$ is shifted by $-\pi/2$ rad (or $-90°$) to generate its Hilbert transform $\cos(\omega t - \pi/2) = \sin(\omega t)$. We may generalize this property and define a

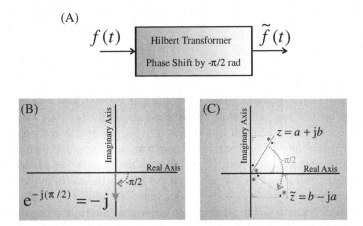

Figure 1.4 (A) The Hilbert transform can be represented as the operation of an LTI system (the Hilbert transformer). Input $f(t)$ is transformed into $\tilde{f}(t)$ by shifting it by $-\pi/2$ rad $(-90°)$. (B) The Hilbert transform operation in the frequency domain can be represented as a multiplication with $e^{-j(\pi/2)} = -j$ (orange arrow). (C) Example of the Hilbert transform in the frequency domain—that is, multiplication of a complex number $z = a + jb$ with $-j$. The result is $\tilde{z} = b - ja$. As can be seen, the result is a $-90°$ rotation. Note in this panel that $90°$ angles are indicated by ∟ and that the angles indicated by • and * add up to $90°$. *(Color in electronic version.)*

Hilbert transformer as a phase-shifting (linear time invariant, LTI) system that generates the Hilbert transform of its input (Fig. 1.4A). The generalization of this property associated with the cosine is not too far of a stretch if you recall that, with the real Fourier series, any periodic signal can be written as the sum of sinusoidal signals (the cosine and sine waves in equation (5.1) in van Drongelen, 2007) and that our above results can be applied to each of these sinusoidal components.

To further define the shifting property of the Hilbert transformer (see Fig. 1.4A), we begin to explore this operation in the frequency domain, because here the procedure of shifting the phase of a signal by $-\pi/2$ rad is relatively easy to define as a multiplication by $e^{-j(\pi/2)} = -j$ (Fig. 1.4B). If this is not obvious to you, consider the effect of this multiplication for any complex number $z = e^{j\phi}$ (representing phase ϕ) that can also be written as the sum of its real and imaginary parts $z = a + jb$. Multiplication by $-j$ gives its Hilbert transform $\tilde{z} = -j(a + jb) = b - ja$, indeed corresponding to a $-90°$ rotation of z (see Fig. 1.4C). Although the multiplication with $-j$ is correct for the positive frequencies, a $-90°$ shift for the negative frequencies in the Fourier transform (due to the negative values of ω) corresponds to multiplication with $e^{j(\pi/2)} = j$. Therefore, the operation of the Hilbert transform in the frequency domain can be summarized as:

$$\boxed{\text{multiplication by} - j\, \text{sgn}(\omega)} \tag{1.11}$$

Here we use the so-called signum function sgn (Appendix 1.2, Fig. A2.1) defined as:

$$
\text{sgn}(\omega)\begin{cases} -1 & \text{for } \omega < 0 \\ 0 & \text{for } \omega = 0 \\ 1 & \text{for } \omega > 0 \end{cases}
\tag{1.12}
$$

Let us go back to our phase-shifting system depicted in Fig. 1.4A and define its unit impulse response as $h(t)$ and its associated frequency response as $H(\omega)$. Within this approach, the Hilbert transform is the convolution of input $f(t)$ with $h(t)$. Using our knowledge about convolution (if you need to review this, see section 8.3.2 in van Drongelen, 2007), we can also represent the Hilbert transform in the frequency domain as the product of $F(\omega)$—the Fourier transform of $f(t)$—and $H(\omega)$. This is very convenient because we just determined above that the Hilbert transform in the frequency domain corresponds to a multiplication with $-j\,\text{sgn}(\omega)$. To summarize, we now have the following three Fourier transform pairs:

System's input $\Leftrightarrow$ Fourier transform: $\qquad\qquad f(t) \Leftrightarrow F(\omega)$
System's unit impulse response $\Leftrightarrow$ Fourier transform: $\quad h(t) \Leftrightarrow H(\omega)$
Hilbert transform $\Leftrightarrow$ Fourier transform: $\qquad\qquad f(t) \otimes h(t) \Leftrightarrow F(\omega)H(\omega)$

$$\tag{1.13}$$

Using these relationships and Equation (1.11), we may state that the Fourier transform of the unit impulse response (i.e., the frequency response) of the Hilbert transformer is:

$$
\boxed{H(\omega) = -j\text{sgn}(\omega)}
\tag{1.14}
$$

We can use the expression we found for $H(\omega)$ to examine the above example of Euler's equation

$$
A\,e^{j\omega_0 t} = \underbrace{A\cos(\omega_0 t)}_{\text{Signal}} + j\underbrace{A\sin(\omega_0 t)}_{\text{Hilbert transform}}
$$

$$\overbrace{\qquad\qquad\qquad\qquad}^{\text{Analytical signal}}$$

in the frequency domain. The Fourier transform of the cosine term (using equation (6.13) in van Drongelen, 2007) is:

$$
A\pi[\delta(\omega + \omega_0) + \delta(\omega - \omega_0)]
\tag{1.15}
$$

Now, according to Equation (1.13), the Fourier transform of the cosine's Hilbert transform is the product of the Fourier transform of the input signal (the cosine) and the frequency response of the Hilbert transformer $H(\omega)$, that is:

$$
\begin{aligned}
&\{A\pi[\delta(\omega+\omega_0)+\delta(\omega-\omega_0)]\}\{-j\,\text{sgn}(\omega)\}\\
&A\pi[\delta(\omega+\omega_0)(-j\,\text{sgn}(\omega)) + \delta(\omega-\omega_0)(-j\,\text{sgn}(\omega))]
\end{aligned}
\tag{1.16}
$$

Because $\delta(\omega + \omega_0)$ is only nonzero for $\omega = -\omega_0$ and $\delta(\omega - \omega_0)$ is only nonzero for $\omega = \omega_0$, we may rewrite the $-j\,\text{sgn}(\omega)$ factors in Equation (1.16) and we get:

$$A\pi[\delta(\omega + \omega_0)(-j\,\text{sgn}(-\omega_0)) + \delta(\omega - \omega_0)(-j\,\text{sgn}(\omega_0))]$$

Now we use the definition of sgn, $\text{sgn}(-\omega_0) = -1$ and $\text{sgn}(\omega_0) = 1$ (Equation (1.12)), and simplify the expression to:

$$A\pi[\delta(\omega + \omega_0)(j) + \delta(\omega - \omega_0)(-j)] = Aj\pi[\delta(\omega + \omega_0) - \delta(\omega - \omega_0)] \qquad (1.17)$$

As expected, Equation (1.17) is the Fourier transform of $A\sin(\omega t)$ (see equation (6.14) in van Drongelen, 2007), which is indeed the Hilbert transform $\tilde{f}(t)$ of $f(t) = A\cos(\omega t)$.

Combining the above results, we can find the Fourier transform of the analytic signal $f_a(t) = A\cos(\omega t) + jA\sin(\omega t) = f(t) + j\tilde{f}(t)$. If we define the following pairs:

$$f_a(t) \Leftrightarrow F_a(\omega)$$
$$f(t) \Leftrightarrow F(\omega)$$
$$\tilde{f}(t) \Leftrightarrow \tilde{F}(\omega)$$

the above expressions can be combined in the following Fourier transform pair:

$$f_a(t) = f(t) + j\tilde{f}(t) \Leftrightarrow F_a(\omega) = F(\omega) + j\tilde{F}(\omega)$$

In the above equation we substitute the expressions for $F(\omega)$ from Equation (1.15) and $\tilde{F}(\omega)$ from Equation (1.17) and get:

$$F_a(\omega) = F(\omega) + j\tilde{F}(\omega)$$

$$= \underbrace{A\pi[\delta(\omega + \omega_0) + \delta(\omega - \omega_0)]}_{\text{Fourier transform of signal}} + j\underbrace{\left\{Aj\pi[\delta(\omega + \omega_0) - \delta(\omega - \omega_0)]\right\}}_{\text{Fourier transform of Hilbert transform}}$$

(overbrace: Fourier transform of analytical signal)

With a bit of algebra we obtain:

$$\boxed{F_a(\omega) = F(\omega) + j\tilde{F}(\omega) = 2\pi A\delta(\omega - \omega_0)} \qquad (1.18)$$

This interesting finding shows that the Fourier transform of the analytic signal has zero energy at negative frequencies and only a peak at $+\omega_0$. The peak's amplitude at $+\omega_0$ is double the size of the corresponding peak in $F(\omega)$ (Fig. 1.3C). This finding may be generalized as: "The Fourier transform of the analytic signal $F_a(\omega)$ has no energy at negative frequencies $-\omega_0$, it only has energy at positive frequencies $+\omega_0$ and its amplitude is double the amplitude at $+\omega_0$ in $F(\omega)$."

1.3.2 The Hilbert Transform in the Time Domain

From the frequency response presented in Equation (1.14) and the relationship between convolution in the time and frequency domains (section 8.3.2, van Drongelen, 2007), we know that the unit impulse response $h(t)$ of the Hilbert transformer (Fig. 1.4A) is the inverse Fourier transform of $-j\,\mathrm{sgn}(\omega)$. You can find details of $\mathrm{sgn}(t)$ and its Fourier transform in Appendix 1.2; using the signum's Fourier transform, we can apply the duality property (section 6.2.1, van Drongelen, 2007) to determine the inverse Fourier transform for $-j\,\mathrm{sgn}(\omega)$. For convenience we restate the duality property as:

$$\text{if } f(t) \Leftrightarrow F(\omega), \text{ then } F(t) \Leftrightarrow 2\pi f(-\omega) \tag{1.19a}$$

Applying this to our signum function (see also Appendix 1.2), we can define the inverse Fourier transform of $\mathrm{sgn}(\omega)$:

$$\mathrm{sgn}(t) \Leftrightarrow \frac{2}{j\omega}, \text{ therefore } \frac{2}{jt} \Leftrightarrow 2\pi \underbrace{\mathrm{sgn}(-\omega)}_{-\mathrm{sgn}(\omega)} \tag{1.19b}$$

Note that we can substitute $-\mathrm{sgn}(\omega) = \mathrm{sgn}(-\omega)$ because the signum function (Fig. A2.1) has odd symmetry (defined in Appendix 5.2 in van Drongelen, 2007). Using the result from applying the duality property in Equation (1.19b), we can determine the inverse Fourier transform for the frequency response of the Hilbert transformer $H(\omega) = -j\,\mathrm{sgn}(\omega)$ and find the corresponding unit impulse response $h(t)$. Because 2π and j are both constants, we can multiply both sides with j and divide by 2π; this generates the following Fourier transform pair:

$$\boxed{h(t) = \frac{1}{\pi t} \Leftrightarrow H(j\omega) = -j\,\mathrm{sgn}(\omega)} \tag{1.20}$$

In Equation (1.14) we found that the frequency response of the Hilbert transformer is $-j\,\mathrm{sgn}(\omega)$. Because we know that multiplication in the frequency domain is equivalent to convolution in the time domain (chapter 8 in van Drongelen, 2007), we can use the result in Equation (1.20) to define the Hilbert transform $\tilde{f}(t)$ of signal $f(t)$ in both the time and frequency domains. We define the following Fourier transform pairs:

the input: $f(t) \Leftrightarrow F(\omega)$
the Hilbert transform of the input: $\tilde{f}(t) \Leftrightarrow \tilde{F}(\omega)$
the unit impulse response of the Hilbert transformer: $h(t) \Leftrightarrow H(\omega)$

Using the above pairs and Equation (1.20), the Hilbert transform and its frequency domain equivalent are:

$$\tilde{f}(t) = f(t) \otimes \underbrace{h(t)}_{\frac{1}{\pi t}} = \frac{1}{\pi} \int_{-\infty}^{\infty} \frac{f(t)}{t - \tau} d\tau \Leftrightarrow \tilde{F}(\omega) = F(\omega)H(\omega) \qquad (1.21)$$

There is, however, a problem with our finding for the Hilbert transform expression in Equation (1.21), which is that there is a pole for $f(t)/(t - \tau)$ within the integration limits at $t = \tau$. The solution to this problem is to define the Hilbert transform as:

$$\boxed{\tilde{f}(t) = \frac{1}{\pi} \text{CPV} \int_{-\infty}^{\infty} \frac{f(t)}{t - \tau} d\tau} \qquad (1.22)$$

in which CPV indicates the Cauchy principal value of the integral. The CPV is a mathematical tool to evaluate integrals that include poles within the integration limits. An example of such an application is given in Appendix 1.3. For those interested in the CPV procedure, we refer to a general mathematics text such as Boas (1966).

1.3.3 Examples

The Hilbert transform is available in MATLAB via the hilbert command. Note that this command produces the analytic signal $f(t) + j\tilde{f}(t)$ and not the Hilbert transform itself; the Hilbert transform is the imaginary component of the output.

You can evaluate the example from Equation (1.10) by computing the Hilbert transform for the cosine and plot the amplitude and phase. Type the following in the MATLAB command window:

```
step=0.00001;                    % step size=1/sample rate
t=0:step:1;                      % timebase
x=cos(2*pi*4*t);                 % 4 Hz signal
xa=hilbert(x);                   % compute the analytic signal
Amplitude=abs(xa);               % amplitude of the signal
Phase=atan2(imag(xa),real(xa));  % instantaneous phase
Ohmega=diff(Phase)/(2*pi*step);  % instantaneous frequency in Hz
figure;plot(t,x,'k');hold;
plot(t,Amplitude,'r');
plot(t,Phase,'g');
plot(t(1:length(t)-1),Ohmega,'m.')
axis([0 1 -5 5])
```

You will obtain a graph of a 4-Hz cosine function with an indication of its amplitude (a constant) in red, its instantaneous phase in green (note that we use the atan2 MATLAB command in the above example because we want to obtain phase angles between $-\pi$ and $+\pi$), and the frequency as the derivative of the phase in magenta.

You can now check both frequency characteristics we discussed by computing the Fourier transforms and plotting these in the same graph.

```
X=fft(x);              % Fourier transform of the signal
XA=fft(xa);            % Fourier transform of the analytic signal
figure;plot(abs(X),'k');hold;plot(abs(XA))
```

If you use the zoom function of the graph to study the peaks in the plot you will see that the peaks for the positive frequencies (far-left part of the graph) show a difference of a factor two between the Fourier transform of the analytic signal and the Fourier transform of the signal. The negative component (in the discrete version of the Fourier transform this is the far-right part of the graph) shows only a peak in the Fourier transform of the signal. Both observations are as expected from the theoretical considerations in Section 1.3.1.

Another property to look at is the phase shift between the signal and its Hilbert transform. This can be accomplished by typing the following lines:

```
figure; hold;
plot(t,imag(xa),'r');      % the imaginary part of the analytic signal=
                           % the Hilbert transform
plot(t,x,'k.')             % the signal
plot(t,real(xa),'y')       % real part of the analytic signal=signal
```

Now you will get a figure with the signal (4-Hz cosine wave) in both black (thick line) and yellow (thin line); the Hilbert transform (the 4-Hz sine wave) is plotted in red.

Finally we will apply these techniques to an example in which we have two neural signals, one signal generated by a single neuron and one signal generated by the network in which the neuron is embedded. Our question here is how the phases of these two signals relate. First, the raw extracellular trace is rectified and sent through a low-pass filter with a 50-ms time constant (this technique of using the analytic signal to find the instantaneous phase usually works better with signals composed of a small band of frequencies and, in our case, we are only interested in the low-frequency behavior; see Pikovsky et al., 2001, for more details). For the cellular activity, we create a raster plot of the spike times and send it through the same low-pass filter. We now have two signals representing the low-pass-filtered spiking behavior of the cell and network (see Fig. 1.5Aiii). We can use the Hilbert

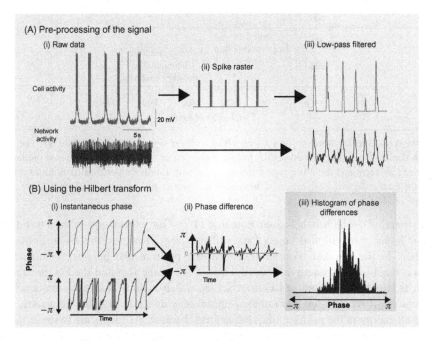

Figure 1.5 (A) Processing of a cellular and network activity (i) into a low-frequency index of spiking activity (iii) (see text for details). (B) The low-pass-filtered signals of (A) were transformed using the analytic signal technique to find the instantaneous phase over time (i). The relationship between the two signals was investigated by finding the difference between the phases over time (ii) and plotting these phase differences in a histogram (iii). In this example we observe that the overall effect is that the network activity leads and the cell activity lags—that is, the histogram (iii) of network activity phase minus cell activity phase is predominantly positive. *(Color in electronic version.) (From A. Martell, unpublished results, with permission.)*

transform technique to find the instantaneous phase of each signal (Fig. 1.5Bi). For our case, we are interested in how the phases of the cellular and network signals are related. To find this relationship, we calculate the difference between the two instantaneous phase signals at each point in time and then use this information to generate a histogram (see Fig. 1.5Bii–iii). This method has been used to compare how the phases of cellular and network signals are related for different types of cellular behavior (Martell et al., 2008).

Appendix 1.1

In the case of the standard power spectrum we have $S = XX^*/N$ (equation (7.1) in van Drongelen, 2007). The normalization by $1/N$ ensures that Parseval's conservation of energy theorem is satisfied (this theorem states that the sum of squares in the time domain and the sum of all elements in the power spectrum are equal; see Table 7.1 and Appendix 7.1 in van Drongelen, 2007). In the case of Lomb's algorithm we compute the sum of squares for each frequency by using the expression in

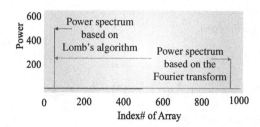

Figure A1.1 Spectral analysis of a 1-s epoch of a 50-Hz signal sampled at 1000 Hz. The graph depicts the superimposed results from a standard power spectrum (red) based on the Fourier transform and the power spectrum obtained with Lomb's algorithm (dark blue). Note that the total energy in both cases is identical. This figure can be created with Pr1_2.m. *(Color in electronic version.)*

Equation (1.8), which is based on Equation (1.7). Our expectation is therefore that Lomb's spectrum will also satisfy Parseval's theorem. However, there is a slight difference. In the standard Fourier transform the positive and negative frequencies each contain half the energy. Basically, this is due to the fact that the Fourier transform is based on the complex Fourier series, which includes negative frequencies. In contrast, if we compute the Lomb spectrum only up to the Nyquist frequency, we have all energy in the positive frequencies, and therefore its values are twice as large as compared to the standard power spectrum. An example for a single frequency is shown in Figure A1.1. This figure is based on a standard power spectrum and Lomb spectrum computed for the same input, a sine wave of 50 Hz. Thus, if we want the Lomb spectrum to have the same amplitudes as the standard power spectrum, we need to divide by two. Furthermore, if we want to normalize by the total power, we can divide by the variance σ^2. This normalization by $2\sigma^2$ is exactly the normalization commonly applied for Lomb's spectrum (see Equation (1.9) and Pr1_1.m).

Appendix 1.2

This appendix describes the signum function sgn(t), its derivative, and Fourier transform. The signum function is 1 for positive t and -1 for negative t (Fig. A2.1). Similar to the derivative of the unit step function $U(t)$ (section 2.2.2, fig. 2.4A in van Drongelen, 2007), the derivative of this function is only nonzero at $t = 0$. The only difference is that for sgn(t) the function increases by 2 units (from -1 to 1) instead of 1 unit (from 0 to 1) in $U(t)$. Since the derivative of the unit step is $\delta(t)$, the derivative of the signum function would be twice as large, that is:

$$\frac{d[\text{sgn}(t)]}{dt} = 2\delta(t) \qquad\qquad (A1.2.1)$$

The Fourier transform of the derivative of a function is equal to the Fourier transform of that function multiplied with jω. This property is similar to the relationship of the Laplace transform of a derivative of a function and the Laplace transform of the function itself (see section 9.3, equation (9.3) in van Drongelen, 2007). If we

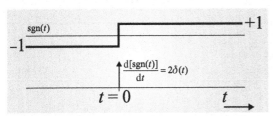

Figure A2.1 The signum function and its derivative, the unit impulse function with an amplitude of two. *(Color in electronic version.)*

now use this property and define the Fourier transform of sgn(t) as $S(\omega)$, we can apply the Fourier transform to both sides of Equation (A1.2.1):

$$j\omega S(\omega) = 2 \qquad (A1.2.2)$$

Recall that the Fourier transform of the unit impulse is 1 (see section 6.2.1, equation (6.9) in van Drongelen, 2007). Therefore, the Fourier transform pair for the signum function is:

$$\text{sgn}(t) \Leftrightarrow S(\omega) = \frac{2}{j\omega} \qquad (A1.2.3)$$

Appendix 1.3

In Equation (1.22) we use the Cauchy principal value, CPV. This technique is used to approach integration of a function that includes a pole within the integration limits. We will not go into the mathematical details (for more on this subject please see a mathematics textbook such as Boas, 1966), but we will give an example to show the principle. For example, consider the integral $\int_{-d}^{d} (1/x)\mathrm{d}x$. The function $1/x$ in this integral has a pole (is unbounded) at $x = 0$. The Cauchy principal value technique approximates the integral as the sum of two separate integral:

$$\int_{-d}^{d} \frac{1}{x}\mathrm{d}x \approx \int_{-d}^{-\varepsilon} \frac{1}{x}\mathrm{d}x + \int_{\varepsilon}^{d} \frac{1}{x}\mathrm{d}x$$

where ε is a small positive value approaching zero. In this case the two integrals cancel and approach $\int_{-d}^{d} (1/x)\mathrm{d}x$. Our final result can be summarized as:

$$\text{CPV} \int_{-d}^{d} \frac{1}{x}\mathrm{d}x = \lim_{\varepsilon \to 0} \left[\int_{-d}^{-\varepsilon} \frac{1}{x}\mathrm{d}x + \int_{\varepsilon}^{d} \frac{1}{x}\mathrm{d}x \right] = 0$$

Here the Cauchy principal value is indicated by CPV; in other texts you may also find PV or P.

2 Modeling

2.1 Introduction

Signal analysis is frequently used to characterize systems. In van Drongelen (2007), chapter 8, we described linear systems and associated techniques that allow us determine system characteristics. In the last chapter of van Drongelen (2007) (section 17.3) we showed how these linear methods, such as cross-correlation, fail to characterize signals with a nonlinear component. To address this shortcoming, we used metrics such as correlation dimension, the Lyapunov exponent, or Kolmogorov entropy to characterize nonlinear signal properties.

The goal of this chapter is to introduce basics for modeling systems, with an emphasis on techniques used to characterize nonlinear systems and their signals. In this context, this chapter will also provide an introduction to the application of the Volterra series, which forms the basis for the identification of dynamical nonlinear systems, and which we will go over in more detail in Chapter 3. The systems that we will introduce in this chapter are considered higher-order systems, since they include operators beyond the (linear) first-order one. Useful references on the characterization of nonlinear systems are the seminal text by Marmarelis and Marmarelis (1978) and the reprint edition of a text from the 1980s by Schetzen (2006). For more recent overviews, see Westwick and Kearney (2003) and Marmarelis (2004).

2.2 Different Types of Models

Before going into mathematical detail, it is useful to summarize some of the types of models that one may encounter in neuroscience. Attenuators and amplifiers are both examples of **linear** systems, since output is simply the product of input and a constant (e.g., $y = 3x$). Alternatively, expressions that characterize **nonlinear** systems include higher-order terms: these systems, as we will see in Chapters 3–5, do not obey the scaling and superposition rules of linear models (to review these rules see section 8.3.1.1 in van Drongelen, 2007). Examples of nonlinear higher-order systems are $y = x^2$ (second-order system) and $y = 5 + x + 3x^3$ (third-order system). At this point it is important to note that an expression such as $y = a + bx + cx^3$ can still be considered linear, but with respect to its parameters a, b, and c. This is a property that we will use when developing the regression procedure in Section 2.4.1.

All of the examples earlier are **static** models (systems without memory), meaning that their output depends only on present input. In neuroscience we usually must deal with **dynamical** models, in which output depends on present and past input (but not on future input); these systems are also called **causal**. Static models are represented by algebraic equations (such as the ones in the previous paragraph), whereas dynamical systems are modeled by differential equations (for continuous time models) or difference equations (for discrete time models). General examples of linear dynamical systems with input x and output y are:

$$A_n \frac{d^n y(t)}{dt^n} + A_{n-1} \frac{d^{n-1} y(t)}{dt^{n-1}} + \cdots + A_0 y(t)$$

$$= B_m \frac{d^m x(t)}{dt^m} + B_{m-1} \frac{d^{m-1} x(t)}{dt^{m-1}} + \cdots + B_0 x(t)$$

for continuous time systems and:

$$A_n y(k - n) + A_{n-1} y(k - n + 1) + \cdots + A_0 y(k)$$

$$= B_m x(k - m) + B_{m-1} x(k - m + 1) + \cdots + B_0 x(k)$$

for discrete time systems (for details see chapter 8 in van Drongelen, 2007). If one of the terms in a differential or difference equation is of a higher order, we have a nonlinear dynamical system. For example, $y - 4(dy/dt)^2 = 2x$ represents a second-order dynamical system.

Time invariance is a critical condition for the development of the convolution formalism (see section 8.3.1.1 in van Drongelen, 2007). This property allows us to state that a system's response to identical stimuli at different points in time is always the same (provided that the system is in the same state, of course). Just as we have linear time invariant systems, we also have nonlinear time invariant systems (usually abbreviated as LTI or NLTI systems).

Models of real systems can be generated according to two major methodological approaches. One might follow a **deductive** path and start from (physical) assumptions about the system, generating a **hypothetical model** to create predictions that can be empirically tested. These empirical measurements can be used to establish the parameters of the hypothetical model, and therefore this type of representation is often called a **parametric model**. An alternative to this procedure, the **inductive** path, is followed if one starts from the measurements of a system's input and output. This data-driven method uses measurements, rather than assumptions about the system, to mathematically relate input and output. Here, we can consider the system as a **black box**, modeled by a mathematical relationship that transforms input into output. This type of model is often referred to as **nonparametric** (note, however, that nonparametric does not refer to the absence of parameters; in many cases, these models will have more parameters than parametric models). The method of induction is appropriate when dealing with complex systems that resist a reduction to a simpler parametric model. It can also be a starting point in which a system is first characterized as a black box and in subsequent steps parts with physical meaning replace pieces of the black box. In this combined approach, parts of the model may still be part of the black box, whereas other parts may be associated with a physical interpretation. In this case, the distinction between parametric and nonparametric models may become a bit fuzzy.

2.3 Examples of Parametric and Nonparametric Models

A **parametric** model usually has relatively few parameters. A simple example of a parametric model of a dynamical LTI system is the ordinary differential equation (ODE) for a filter. For example, $x = RC(dy/dt) + y$ describes input x and output y of a simple RC circuit (Fig. 2.1A). The only parameters in this case are the values of the resistor R and the capacitor C in the equation. Subsequently, the value for these parameters can be determined experimentally from observing the system's behavior.

Note: See van Drongelen (2007) for further details about determining these parameters from measurements: in section 11.2.1 it is shown how RC can be obtained from the filter's unit step response (equation (11.8)), and in section 12.3, RC is determined from the -3 dB point of the filter's frequency characteristic (equation (12.5)).

A very famous parametric model in neuroscience is the Hodgkin and Huxley (1952) formalism using four variables to describe the action potential generated in the squid's giant axon: the membrane potential V and three other variables m, h, and n describe the membrane potential-dependent characteristics of sodium and potassium conductivity. In the following it is assumed you are somewhat familiar with the Hodgkin and Huxley model; if you need to review the details, chapter 2 in Izhikevich (2007) provides an excellent overview.

Initially in the 1950s, the formalism was entirely hypothetical, and it was not until after the molecular basis for Na^+ and K^+ ion channels was elucidated that a physical interpretation of large parts of the model could be made. The gating variable m characterizes the depolarization process caused by increased conductance of sodium ions (causing an influx of positively charged sodium ions) that occurs during the action potential generation. The variables h and n are recovery variables that represent the inactivation of sodium ion conductance (reduced Na^+ influx) and the activation of potassium conductance (causing an outward flux of positively charged potassium ions).

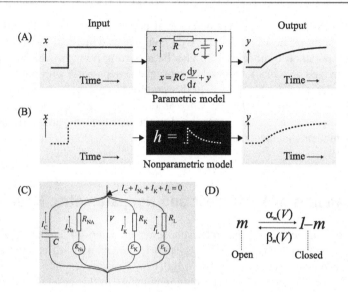

Figure 2.1 (A) Example of a parametric model of a dynamical linear system (a low-pass filter) and its input and output (*x* and *y* respectively). (B) The black box, nonparametric equivalent of the same system is the white curve representing the (sampled) unit impulse response (UIR). Both models permit us to predict the output resulting from an arbitrary input such as the unit step function. The parametric model has two parameters (*R* and *C*) with physical meaning. The nonparametric model consists of many parameters (the samples making up the UIR) without a direct physical meaning. (C) Hodgkin and Huxley's electronic equivalent circuit for the biomembrane. The model consists of the membrane capacitance (*C*) and three parallel ion channels: one for sodium, one for potassium, and a leakage channel. According to Kirchhoff's first law the sum of all currents at the node (*arrow*) must be zero. (D) Model for gating variable *m* in the Hodgkin and Huxley formalism. *(Color in electronic version.)*

Hodgkin and Huxley's model relates all these variables in an equivalent circuit of the excitable biomembrane (Fig. 2.1C) by setting the sum of all membrane currents equal to zero according to Kirchhoff's first law (see appendix 1.1 in van Drongelen, 2007). By applying this law to the node indicated with the arrow in the membrane model in Fig. 2.1C we obtain:

$$C\frac{dV}{dt} + \frac{V - E_{Na}}{R_{Na}} + \frac{V - E_K}{R_K} + I_L = 0 \tag{2.1}$$

Capacitive current $= I_C$; Sodium current $= I_{Na}$; Potassium current $= I_K$; Leak current

In this expression we have several parameters: membrane capacitance *C*; the resistance values for sodium and potassium ions, R_{Na} and R_K, respectively; E_{Na} and E_K

are the equilibrium potentials for sodium and potassium ions computed with the Nernst equation (appendix 1.1 in van Drongelen, 2007); and I_L is a constant leakage current attributed to Cl^- ions. The sodium and potassium currents are determined with Ohm's law (appendix 1.1 in van Drongelen, 2007): each ion species experiences a potential drop equal to the difference between the membrane potential V and its equilibrium potential (e.g., for sodium: $V - E_{Na}$), and this potential drop divided by the resistance is the ion current (e.g., for sodium the current is $(V - E_{Na})/R_{Na}$). In addition to Equation (2.1), Hodgkin and Huxley (1952) described the dynamics for R_{Na} and R_K with the nonlinear relationships $g_{Na} = 1/R_{Na} = \bar{g}_{Na}m^3h$ and $g_K = 1/R_K = \bar{g}_K n^4$, where $\bar{g}_{Na}$ and $\bar{g}_K$ are the maximum conductivity values for sodium and potassium. Furthermore, the gating variable m is modeled by a reversible process between the open (m) and closed ($1-m$) states (Fig. 2.1D), which can be represented by the following ODE:

$$\frac{dm}{dt} = \alpha_m(V)(1 - m) - \beta_m(V)m \tag{2.2}$$

The rate parameters α_m and β_m that govern this process depend on the membrane potential V in a nonlinear fashion. The two other gating variables h and n follow the same formalism with membrane potential-dependent rate constants α_h, β_h, α_n, and β_n. Hodgkin and Huxley determined these nonlinear relationships between the rate parameters and membrane potential from voltage clamp experiments.

Over time, other ion channels were introduced into the model using the same formalism as for the sodium and potassium channels. Since the development of computer technology, the Hodgkin and Huxley formalism has been widely used in simulations of neural systems ranging from very detailed models of single neurons (e.g., De Schutter and Bower, 1994a, b) to large-scale networks of neocortex (e.g., Traub et al., 2005; van Drongelen et al., 2006).

Although Hodgkin and Huxley's model only contains four variables (V, m, h, n), it is still too complex to approach analytically. Several authors solved this problem by reducing the 4D model to a 2D one; the Fitzhugh−Nagumo model (Fitzhugh, 1961) is an example of such a reduction. In these models, the gating variable m of the Hodgkin and Huxley model is removed by considering sodium activation to be instantaneous; subsequently, h and n are combined into a single recovery variable w. Fitzhugh used the following pair of coupled differential equations:

$$\frac{dV}{dt} = V(a - V)(V - 1) - w + I \quad \text{and} \quad \frac{dw}{dt} = bV - cw \tag{2.3}$$

in which a, b, and c are parameters; I is a term representing injected current. The equations are coupled because w occurs in the expression for dV/dt and V in the expression for dw/dt. The remaining two variables in these models are the membrane potential V and a single recovery variable w, generating a 2D model that is amenable to mathematical analysis (for an excellent discussion of simplified versions of the Hodgkin and Huxley model, see Izhikevich, 2007).

Nonparametric models describe a system's input−output relationship, usually by using a large number of parameters, and these parameters do not necessarily have a physical interpretation. Generally speaking, a nonparametric model is generated from a procedure in which we relate a system's input $x(t)$ and output $y(t)$. Just as we can relate two variables with a function, we can link two time series with an **operator**. An example of such a nonparametric model would be the characterization of an LTI dynamical system with its (sampled) unit impulse response (UIR) (Fig. 2.1B). The operator in this case would be convolution, since convolution of the input time series $x(t)$ with the system's UIR $h(t)$ generates the system's output time series $y(t)$: $y(t) = h(t) \otimes x(t)$ (see section 8.3.1.1 in van Drongelen, 2007). Although one might point out that such a nonparametric description does not necessarily provide direct insight into the system's components or the mechanisms underlying the system's operation, the curve of the UIR permits us to predict the system's response to any input, such as the unit step function (Fig. 2.1B).

2.4   Polynomials

For static systems, both linear and nonlinear, one can use algebraic expressions to describe their input−output characteristic, and polynomials are often used for this purpose. Polynomials are sums of monomials, which are expressions that consist of a constant multiplied by one or more variables; the exponent of the variable is its degree. For example, $z(t) = ax(t)^4 y(t)^3$ is a monomial with a constant (parameter) a and a degree of 4 for x and 3 for y. We can see that this expression represents a static process because at any time t, output z depends only on the present values of inputs x and y. It is important to note that although the relationship between z and x, y is nonlinear, the expression can be considered a linear function of the parameter a.

2.4.1   Describing Discrete Time Data Sets

Applying the above to the characterization of nonlinear systems, we could describe the relationship between input $x(t)$ and output $y(t)$ of a static nonlinearity (a nonlinear system without memory) with a polynomial such as the following power series:

$$y(t) = a_0 + a_1 x(t) + a_2 x(t)^2 + a_3 x(t)^3 + \cdots + a_i x(t)^i + \cdots = \sum_{i=0}^{\infty} a_i x(t)^i \qquad (2.4)$$

In principle, power series are infinite; however, in our applications they will always consist of a finite number of monomials. The fact that Equation (2.4) **is linear with respect to its parameters** a_i can be used to fit the series by using a technique called **least squares** minimization. Using this approach of fitting polynomials to recorded data sets is often called **regression analysis**. This procedure works as follows. Suppose we have two sets of N measurements: a system's input x_n and associated output y_n. If we model our system as a second-order static system, we can truncate the expression in Equation (2.4) above the second power and estimate the output y as $a_0 + a_1 x_n + a_2 x_n^2$. Subsequently we can define the error of our estimate ε^2 as:

$$\varepsilon^2 = \sum_{n=1}^{N} [y_n - (a_0 + a_1 x_n + a_2 x_n^2)]^2 \tag{2.5}$$

By following the same approach we used to find the coefficients in Lomb's algorithm (Section 1.2.1), we can find the minimum associated with the best choice for parameters a_0, a_1, and a_2 by setting the partial derivatives of ε^2 (with respect to these three parameters a_0, a_1, and a_2) equal to zero:

$$\frac{\partial \varepsilon^2}{\partial a_i} = \sum_{n=1}^{N} 2 [y_n - (a_0 + a_1 x_n + a_2 x_n^2)] \frac{\partial [y_n - (a_0 + a_1 x_n + a_2 x_n^2)]}{\partial a_i} = 0 \tag{2.6a}$$
$$\text{for } i = 0, 1, 2$$

and we get what are called the **normal equations**:

$$\frac{\partial \varepsilon^2}{\partial a_0} = -2 \sum_{n=1}^{N} [y_n - a_0 - a_1 x_n - a_2 x_n^2] = 0$$

$$\rightarrow a_0 \underbrace{N}_{\sum_{n=1}^{N} 1} + a_1 \sum_{n=1}^{N} x_n + a_2 \sum_{n=1}^{N} x_n^2 = \sum_{n=1}^{N} y_n$$

$$\frac{\partial \varepsilon^2}{\partial a_1} = -2 \sum_{n=1}^{N} [y_n - a_0 - a_1 x_n - a_2 x_n^2] x_n = 0$$

$$\rightarrow a_0 \sum_{n=1}^{N} x_n + a_1 \sum_{n=1}^{N} x_n^2 + a_2 \sum_{n=1}^{N} x_n^3 = \sum_{n=1}^{N} y_n x_n$$

$$\frac{\partial \varepsilon^2}{\partial a_2} = -2 \sum_{n=1}^{N} [y_n - a_0 - a_1 x_n - a_2 x_n^2] x_n^2 = 0$$

$$\rightarrow a_0 \sum_{n=1}^{N} x_n^2 + a_1 \sum_{n=1}^{N} x_n^3 + a_2 \sum_{n=1}^{N} x_n^4 = \sum_{n=1}^{N} y_n x_n^2 \tag{2.6b}$$

Note that in Equation (2.6b) all summation (Σ) expressions are numbers that can be computed from the observations; therefore, there are three linear equations with three unknown parameters a_0, a_1, and a_2 to compute (this should be no problem provided, of course, that the set of equations can be solved). Note that if we had truncated Equation (2.4) at a_1, the normal equations that we would have obtained would have been the well-known equations for linear regression.

It is a bit tedious to solve the three equations in (2.6b); therefore, one might prefer to solve the coefficients by using the matrix notation $XA = Y$ for the three equations:

$$
\underbrace{\begin{bmatrix} N & \sum\limits_{n=1}^{N} x_n & \sum\limits_{n=1}^{N} x_n^2 \\ \sum\limits_{n=1}^{N} x_n & \sum\limits_{n=1}^{N} x_n^2 & \sum\limits_{n=1}^{N} x_n^3 \\ \sum\limits_{n=1}^{N} x_n^2 & \sum\limits_{n=1}^{N} x_n^3 & \sum\limits_{n=1}^{N} x_n^4 \end{bmatrix}}_{X} \underbrace{\begin{bmatrix} a_0 \\ a_1 \\ a_2 \end{bmatrix}}_{A} = \underbrace{\begin{bmatrix} \sum\limits_{n=1}^{N} y_n \\ \sum\limits_{n=1}^{N} y_n x_n \\ \sum\limits_{n=1}^{N} y_n x_n^2 \end{bmatrix}}_{Y}
\tag{2.6c}
$$

The coefficients can be found by solving (Equation (2.6c)): that is, $A = X^{-1}Y$. In MATLAB we can use the $\backslash$ operator to obtain this result: $A = X\backslash Y$. An example ($Pr2\_1.m$) for approximating an exponential function $y = e^x$ is available on http://www.elsevierdirect.com/companions/9780123849151.

2.4.2 Describing Analytic Functions

The previous example works with discrete time data such as a set of digitized recordings of a system's input and output. In other applications, one might deal with a parametric model and consequently have access to analytic functions that describe some nonlinear system under investigation (recall that an **analytic function** is smooth and differentiable and that this is not the same as the analytic signal we introduced for the Hilbert transform in Chapter 1). In this case, the so-called Maclaurin or Taylor series approaches, which will be explained in Sections 2.4.2.1 and 2.4.2.2, may be applied to convert the function into a power series. Such a power series approach can also be helpful for creating a linear approximation of a nonlinear function in the neighborhood of a point of interest. Because linear relationships are easier to analyze than nonlinear ones, this technique of linearization of nonlinear functions can help us understand the behavior of complex nonlinear processes.

Like the polynomials discussed in the previous section, the Maclaurin and Taylor series describe static systems. To describe dynamical systems, we can use the Volterra series, which is discussed in detail in Chapter 3. In Section 2.5, we will show that the Taylor series can be considered the static version of a Volterra series.

2.4.2.1 Maclaurin Series

A famous power series describing a function about the origin is the **Maclaurin** series. Let us consider an example with the exponential function and use the power series approach in Equation (2.4) to represent this function:

$$
f(t) = e^t = a_0 + a_1 t + a_2 t^2 + a_3 t^3 + \cdots + a_i t^i + \cdots = \sum_{i=0}^{\infty} a_i t^i
\tag{2.7}
$$

The task at hand is to determine the values of the coefficients a_i for function e^t. We can use the following approach to perform this task. First we determine the derivatives of f.

$$f(t) = e^t = a_0 + a_1 t + a_2 t^2 + a_3 t^3 + \cdots + a_i t^i + \cdots$$

$$\frac{df(t)}{dt} = e^t = a_1 + 2a_2 t + 3a_3 t^2 + \cdots + ia_i t^{i-1} + \cdots$$

$$\frac{d^2 f(t)}{dt^2} = e^t = 2a_2 + (2 \times 3)a_3 t + \cdots + (i \times (i-1))a_i t^{i-2} + \cdots \qquad (2.8)$$

$$\frac{d^3 f(t)}{dt^3} = e^t = (2 \times 3)a_3 + \cdots + (i \times (i-1) \times (i-2))a_i t^{i-3} + \cdots$$

$$\vdots$$

The second step is to consider $f(t) = e^t$ about the origin. As we approach the origin (i.e., t becomes 0), Equation (2.8) simplifies to:

$$f(0) = e^0 = 1 = [a_0 + a_1 t + a_2 t^2 + a_3 t^3 + \cdots + a_i t^i + \cdots]_{t=0} = a_0$$

$$\frac{df(0)}{dt} = e^0 = 1 = [a_1 + 2a_2 t + 3a_3 t^2 + \cdots + ia_i t^{i-1} + \cdots]_{t=0} = a_1$$

$$\frac{d^2 f(0)}{dt^2} = e^0 = 1 = [2a_2 + (2 \times 3)a_3 t + \cdots + (i \times (i-1))a_i t^{i-2} + \cdots]_{t=0} = 2a_2 \qquad (2.9)$$

$$\frac{d^3 f(0)}{dt^3} = e^0 = 1 = [(2 \times 3)a_3 + \cdots + (i \times (i-1) \times (i-2))a_i t^{i-3} + \cdots]_{t=0}$$

$$= (2 \times 3)a_3$$

$$\vdots$$

With the results obtained in Equation (2.9), we can see that for the function, the values for the coefficients a_i are:

$$a_i = \frac{1}{i!} \qquad (2.10)$$

Combining this result in Equation (2.10) with Equation (2.7), we have found the well-known power series expansion of the exponential function:

$$f(t) = e^t = 1 + \frac{1}{1!}t + \frac{1}{2!}t^2 + \frac{1}{3!}t^3 + \cdots + \frac{1}{i!}t^i + \cdots = \sum_{i=0}^{\infty} \frac{1}{i!}t^i \qquad (2.11)$$

In the last expression $\sum_{i=0}^{\infty}(1/i!)\,t^i$, we use the definition $0! \equiv 1$. Note that by using this approach, we include only values of t—there are no previous or future values $(t \pm \tau)$ included in the power series; therefore, this approach is **static** (or memoryless). An example of this approximation is implemented in MATLAB script Pr2_1.m (http://www.elsevierdirect.com/companions/9780123849151).

In the above example we used the exponential $\exp(t)$ for $f(t)$; if we consider the development of Equation (2.4) for any function that can be differentiated, we get:

$$f(0) = [a_0 + a_1 t + a_2 t^2 + a_3 t^3 + \cdots + a_i t^i + \cdots]_{t=0} = a_0 \to a_0 = f(0)$$

$$\frac{df(0)}{dt} = [a_1 + 2a_2 t + 3a_3 t^2 + \cdots + i a_i t^{i-1} + \cdots]_{t=0} = a_1 \to a_1 = f'(0)$$

$$\frac{d^2 f(0)}{dt^2} = [2a_2 + (2 \times 3)a_3 t + \cdots + (i \times (i-1))a_i t^{i-2} + \cdots]_{t=0} = 2a_2 \to a_2 = \frac{f''(0)}{2}$$

$$\frac{d^3 f(0)}{dt^3} = [(2 \times 3)a_3 + \cdots + (i \times (i-1) \times (i-2))a_i t^{i-3} + \cdots]_{t=0}$$

$$= (2 \times 3)a_3 \to a_3 = \frac{f'''(0)}{(2 \times 3)}$$

$$\vdots$$

(2.12)

Here the notation $f'(0)$, $f''(0)$, $f'''(0)$, ... are not functions but represent the numbers computed as the value of the 1st, 2nd, 3rd, ... derivatives of f at $t = 0$. From this more general notation we obtain the expression for the so-called Maclaurin series of $f(t)$:

$$\boxed{f(t) = f(0) + \frac{1}{1!} t f'(0) + \frac{1}{2!} t^2 f''(0) + \frac{1}{3!} t^3 f'''(0) + \cdots}$$

(2.13)

2.4.2.2 Taylor Series

In the above example, we developed the power series for a function about the origin. The development of the Taylor series follows a similar approach but now about any point α. For a power series of power N this becomes:

$$f(t) = a_0 + a_1(t - \alpha) + a_2(t - \alpha)^2 + a_3(t - \alpha)^3 + \cdots + a_i(t - \alpha)^i + \cdots$$

$$= \sum_{i=0}^{\infty} a_i(t - \alpha)^i$$

(2.14)

We will now use a similar approach for the development of this series about α as we used in the Maclaurin series about the origin—except in this case we set $t = \alpha$

(instead of $t = 0$) so that all terms in Equation (2.14) with $(t-\alpha)^i$ vanish. By following this procedure we get:

$$f(\alpha) = [a_0 + a_1(t-\alpha) + a_2(t-\alpha)^2 + a_3(t-\alpha)^3 + \cdots + a_i(t-\alpha)^i + \cdots]_{t=\alpha}$$

$$= a_0 \rightarrow a_0 = f(\alpha)$$

$$\frac{df(\alpha)}{dt} = [a_1 + 2a_2(t-\alpha) + 3a_3(t-\alpha)^2 + \cdots + ia_i(t-\alpha)^{i-1} + \cdots]_{t=\alpha}$$

$$= a_1 \rightarrow a_1 = f'(\alpha)$$

$$\frac{d^2f(\alpha)}{dt^2} = [2a_2 + (2 \times 3)a_3(t-\alpha) + \cdots + (i \times (i-1))a_i(t-\alpha)^{i-2} + \cdots]_{t=\alpha}$$

$$= 2a_2 \rightarrow a_2 = \frac{f''(\alpha)}{2}$$

$$\frac{d^3f(\alpha)}{dt^3} = [(2 \times 3)a_3 + \cdots + (i \times (i-1) \times (i-2))a_i(t-\alpha)^{i-3} + \cdots]_{t=0}$$

$$= (2 \times 3)a_3 \rightarrow a_3 = \frac{f'''(\alpha)}{(2 \times 3)}$$

(2.15)

$$\vdots$$

Similar to the notation used in the previous section, the notation $f'(\alpha)$, $f''(\alpha)$, $f'''(\alpha)$, ... does not refer to functions, but represents the numbers computed as the value of the 1st, 2nd, 3rd, ... derivatives of f at $t = \alpha$. Substituting the findings in Equation (2.15) into Equation (2.14) we obtain the Taylor series about $t = \alpha$:

$$\boxed{f(t) = f(\alpha) + \frac{1}{1!}(t-\alpha)f'(\alpha) + \frac{1}{2!}(t-\alpha)^2 f''(\alpha) + \frac{1}{3!}(t-\alpha)^3 f'''(\alpha) + \cdots}$$ (2.16)

Comparing Equations (2.13) and (2.16), we can establish that the Maclaurin series is the same as a Taylor series computed about the origin (i.e., $\alpha = 0$). This approach can be extended to higher-dimensional systems with multiple inputs; see Appendix 2.1 for examples of the 2D case. It must be noted that in many texts the distinction between Maclaurin and Taylor series is not always made and it is not uncommon to use the term Taylor series for both, a habit we will adopt in the following.

The number of terms in a Taylor series may be infinite. However, if we evaluate a system close to an equilibrium at the origin or α, the value of t or $(t-\alpha)$ is a small number $\ll 1$; therefore, one can ignore higher-order terms in the power series t^n or $(t-\alpha)^n$ because they become increasingly smaller. Thus, in general we can approximate any function close to α with a linear expression obtained from a Taylor series in which higher-order terms are ignored $f(t) \approx f(\alpha) + (t-\alpha)f'(\alpha)$; or, in the case where we evaluate the expression about the origin, we obtain the approximation $f(t) \approx f(0) + tf'(0)$. This technique of **linearizing a nonlinear function** plays an important role in the analysis of nonlinear systems. A system's behavior in a restricted part of its domain can be understood and approximated by a linear version of its characteristic. Sometimes, with the more complex systems, a piecewise approximation with linear functions is the best option for their analysis. For example, if we wanted to evaluate $\sin(t)$ around the origin, we can apply Equation (2.13) and find the series:

$$\sin(t) = \underbrace{\sin(0)}_{0} + \underbrace{t\cos(0)}_{1} - \frac{t^2}{2}\underbrace{\sin(0)}_{0} - \frac{t^3}{6}\underbrace{\cos(0)}_{1} + \cdots$$

For small values of t (around 0) we may ignore all higher-order terms and we find that $\sin(t) \approx t$. In general, such an approach may be useful if one studies a system close to an equilibrium. For example, if one examines a neuron's subthreshold behavior, one must describe the membrane potential close to the resting potential; in this case it makes sense to linearize the nonlinear equations that govern the cell's electrical activity around resting potential. An example of this approach, where the nonlinear Hodgkin and Huxley equations are linearized, can be found in Chapter 10 in Koch (1999).

When fitting a truncated power series to an analytic function, one could truncate the Taylor series at the desired order. However, due to the error introduced by truncation, one may actually obtain a better fit by using a linear regression approach. An example is if one wants to approximate e^t with a second-order power function over a limited interval. The truncated Taylor series (see Equation (2.11)) is $1 + t + 0.5t^2$, but with a regression approach over the interval $[-1,1]$ one obtains a better fit with $0.9963 + 1.1037t + 0.5368t^2$. This can be seen by running MATLAB script Pr2_1 (available on http://www.elsevierdirect.com/companions/9780123849151) where the original exponential function (red), the Taylor series (blue), and the regression result (black) are superimposed. The regression approach for obtaining a power series approximation is also a valid solution if the Taylor series cannot be applied, as in the case of a function that is nonanalytic, such as $y = |x|$ (no [unique] derivative at $x = 0$).

2.5 Nonlinear Systems with Memory

In the above examples, the output $y(t)$ of the nonlinear systems could be described with a polynomial of $x(t)$ because there was a direct relationship between x and y; that is, in these examples there was no memory in the system. However, nonlinear systems with memory do exist, and for these systems we must describe how the output $y(t)$ depends on both the present and the past input: $x(t-\tau)$ with $\tau \geq 0$.

In the following chapter, we will consider the details of the so-called Volterra series for the characterization of dynamical nonlinear systems (nonlinear systems that do have a memory). Here we will demonstrate the similarities between the Volterra and Taylor series. With the Taylor series we can link output value $y = f(x)$ to input value x in the following manner:

$$y = f(\alpha) + \frac{1}{1!}(x - \alpha)f'(\alpha) + \frac{1}{2!}(x-\alpha)^2 f''(\alpha) + \frac{1}{3!}(x-\alpha)^3 f'''(\alpha) + \cdots \quad (2.17)$$

In the example below, we will approximate a nonlinearity with a series truncated at the second order:

$$y(t) = a_0 + a_1 x(t) + a_2 x(t)^2 \qquad (2.18)$$

Before we introduce the Volterra series, we generalize the procedure in which we relate two **values** x and y into a slightly altered procedure in which we relate a pair of **time series** $x(t)$ and $y(t)$. Just as we can relate two values x and y with a **function** f:

$$y = f(x) \qquad (2.19a)$$

we can link two time series $x(t)$ and $y(t)$ with an **operator** F:

$$y(t) = F\{x(t)\} \qquad (2.19b)$$

Note: In some texts on Volterra series F will be called a **functional**. Because F connects two functions $x(t)$ and $y(t)$, it is better to use the term "operator" because strictly speaking, a **functional** maps a function onto a value, whereas an **operator** maps one function to another function.

A Volterra series can perform such an operation:

$$y(t) = \underbrace{h_0}_{\text{0th order term}} + \underbrace{\int_{-\infty}^{\infty} h_1(\tau_1) x(t - \tau_1) d\tau_1}_{\text{1st order term}}$$

$$+ \underbrace{\int_{-\infty}^{\infty} \int_{-\infty}^{\infty} h_2(\tau_1, \tau_2) x(t - \tau_1) x(t - \tau_2) d\tau_1 \, d\tau_2}_{\text{2nd order term}} + \cdots$$

$$+ \underbrace{\int_{-\infty}^{\infty} \int_{-\infty}^{\infty} \cdots \int_{-\infty}^{\infty} h_n(\tau_1, \tau_2, \cdots, \tau_n) x(t - \tau_1) x(t - \tau_2) \cdots x(t - \tau_n) d\tau_1 \, d\tau_2 \cdots d\tau_n}_{n\text{th order term}} \qquad (2.20)$$

Do not be intimidated by this first appearance of the expression for the Volterra series. In the following text and Chapter 3 we will discuss and explain the different components of this representation. At this point it is worthwhile to mention that the Volterra series is essentially the convolution integral extended to nonlinear systems. We could simplify the notation in Equation (2.20) with the commonly used symbol for convolution $\otimes$ (chapter 8 in van Drongelen, 2007), and we get:

$$y = h_0 + h_1 \otimes x + h_2 \otimes x \otimes x + \cdots + h_n \underbrace{\otimes x \otimes \cdots \otimes x}_{n \text{ copies of } x}$$

In the Volterra series (Equation 2.20), input function $x(t)$ determines the output function $y(t)$. The expression is analogous to the Taylor series except that the differentials of the Taylor series are replaced by integrals. The symbols h_0, h_1, h_2, and h_n represent the so-called Volterra kernels. The term "kernel" is uniquely defined for this type of series and should not be confused with the use of this term in computer science or other areas in mathematics. Note that the first-order component $\int_{-\infty}^{\infty} h_1(\tau_1)x(t - \tau_1)d\tau_1$ in the Volterra series is the convolution integral (see section 8.3.1.1 in van Drongelen, 2007) and the higher-order components in Equation (2.20) are convolution-like integrals. Thus for a linear system, kernel h_1 is the UIR. Representations that utilize Volterra series are usually nonparametric—that is, one can predict system output when the input is known, but one cannot necessarily intuit the system's components or underlying mechanisms. In the following we will examine examples of the relationship between Volterra and Taylor series. See also Chapter 3 for further details on the Volterra series.

Despite the similarities between the Taylor series in Equation (2.17) and the Volterra series in Equation (2.20) discussed above, it may not be immediately obvious that they are related. Therefore, we will discuss the similarities for a simple dynamical nonlinear system, which we will subsequently transform into a static nonlinear one. Let us consider a dynamical second-order system that consists of a cascade of a dynamical linear component and a static nonlinear module (Fig. 2.2A). Such a cascade approach with the dynamics in the linear component combined with static nonlinearities is frequently applied in dynamical nonlinear system analysis. In this example, we have the linear component's UIR $h(t)$ and the static second-order nonlinear component $a_0 + a_1 y + a_2 y^2$ (Equation (2.18)). From Fig. 2.2A we can establish that the output y of the linear module can be obtained from the convolution of the input x and the linear module's UIR h:

$$y(t) = \int_{-\infty}^{\infty} h(\tau)x(t - \tau)d\tau \tag{2.21}$$

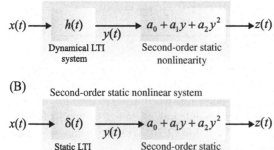

(A) Second-order dynamical nonlinear system

$x(t) \longrightarrow$ | $h(t)$ | $\xrightarrow{y(t)}$ | $a_0 + a_1 y + a_2 y^2$ | $\longrightarrow z(t)$

Dynamical LTI system

Second-order static nonlinearity

(B) Second-order static nonlinear system

$x(t) \longrightarrow$ | $\delta(t)$ | $\xrightarrow{y(t)}$ | $a_0 + a_1 y + a_2 y^2$ | $\longrightarrow z(t)$

Static LTI system

Second-order static nonlinearity

Figure 2.2 (A) Diagram of a second-order dynamical nonlinear system consisting of a cascade of a dynamical LTI system and a second-order static nonlinearity. (B) A similar system for which the dynamical linear component is replaced by a static one. *(Color in electronic version.)*

The cascade's final output z can be obtained from the static nonlinearity characteristic by substituting the output of the linear component (Equation (2.21)) into the
input of the static nonlinearity (Equation (2.18)):

$$z(t) = a_0 + a_1 \int_{-\infty}^{\infty} h(\tau)x(t-\tau)\mathrm{d}\tau + a_2 \left[\int_{-\infty}^{\infty} h(\tau)x(t-\tau)\mathrm{d}\tau \right]^2 \qquad (2.22)$$

This can be rewritten as:

$$z(t) = a_0 + a_1 \int_{-\infty}^{\infty} h(\tau)x(t-\tau)\mathrm{d}\tau$$

$$+ a_2 \underbrace{\left[\left(\int_{-\infty}^{\infty} h(\tau_1)x(t-\tau_1)\mathrm{d}\tau_1 \right) \left(\int_{-\infty}^{\infty} h(\tau_2)x(t-\tau_2)\mathrm{d}\tau_2 \right) \right]}_{\displaystyle \int_{-\infty}^{\infty}\int_{-\infty}^{\infty} h(\tau_1)h(\tau_2)x(t-\tau_1)x(t-\tau_2)\mathrm{d}\tau_1\,\mathrm{d}\tau_2} \qquad (2.23)$$

This expression can be rearranged in the form of the Volterra series shown in
Equation (2.20):

$$z(t) = \overbrace{\underbrace{a_0}_{h_0}}^{\text{0th order term}} + \overbrace{\int_{-\infty}^{\infty} \underbrace{a_1 h(\tau)}_{h_1(\tau)}\, x(t-\tau)\mathrm{d}\tau}^{\text{1st order term}}$$

$$+ \overbrace{\int_{-\infty}^{\infty}\int_{-\infty}^{\infty} \underbrace{a_2 h(\tau_1)h(\tau_2)}_{h_2(\tau_1,\tau_2)}\, x(t-\tau_1)x(t-\tau_2)\mathrm{d}\tau_1\,\mathrm{d}\tau_2}^{\text{2nd order term}} \qquad (2.24)$$

Equation (2.24) shows that the system in Fig. 2.2A can be characterized by a
Volterra series for a second-order system with Volterra kernels h_0, h_1, and h_2.

To demonstrate that the Taylor series is the static equivalent of the Volterra
series, we show the equivalence of Equation (2.24) to the power series in Equation
(2.18). To accomplish this, we consider the case where our dynamical component
in the cascade becomes static; the linear component is now replaced by the static
function $y(t) = x(t)$. In other words, the linear module's UIR is the unit impulse δ

itself, indicating that for this linear component output equals input (Fig. 2.2B). Therefore, we can substitute $\delta(t)$ for $h(t)$ in Equation (2.24):

$$z(t) = a_0 + a_1 \underbrace{\int_{-\infty}^{\infty} \delta(\tau)x(t-\tau)d\tau}_{x(t)}$$

$$+ a_2 \underbrace{\int_{-\infty}^{\infty}\int_{-\infty}^{\infty} \delta(\tau_1)\delta(\tau_2)x(t-\tau_1)x(t-\tau_2)d\tau_1\,d\tau_2}_{x(t)x(t)=x(t)^2} \qquad (2.25)$$

$$= a_0 + a_1x(t) + a_2x(t)^2$$

Thus, in the static case, we can use the Volterra series to recover $z(t) = a_0 + a_1x(t) + a_2x(t)^2$, which is the original expression of the power series in Equation (2.18).

> *Note*: The integrals in Equation (2.25) are evaluated using the property of the unit impulse δ (see section 2.2.2 in van Drongelen, 2007): $\int_{-\infty}^{\infty} x(\tau)\delta(\tau)d\tau = x(0)$ and accordingly $\int_{-\infty}^{\infty} x(t-\tau)\delta(\tau)d\tau = x(t)$.

Appendix 2.1

Taylor Series for a 2D Function

We can extend the Taylor series in Equation (2.16) to a function $f(\tau, \sigma)$ of two variables τ and σ. In the case where we can subdivide the function into two separate ones (e.g., $f(\tau,\sigma)=f(\tau)+f(\sigma)$ or $f(\tau,\sigma)=f(\tau)f(\sigma)$), we can compute the Taylor series for each function $f(\tau)$ and $f(\sigma)$ and add or multiply the individual series to obtain the expression for $f(\tau,\sigma)$. Such an approach would work if, for example, $f(\tau,\sigma)=e^{\tau}\sin(\sigma)$.

Alternatively, one can approach the development of a 2D Taylor series more generally, and consider f about point α, β.

$$f(\tau,\sigma) = a_{00} + a_{10}(\tau-\alpha) + a_{01}(\sigma-\beta) + a_{20}(\tau-\alpha)^2 + a_{11}(\tau-\alpha)(\sigma-\beta)$$

$$+ a_{02}(\sigma-\beta)^2 + a_{30}(\tau-\alpha)^3 + a_{21}(\tau-\alpha)^2(\sigma-\beta)$$

$$+ a_{12}(\tau-\alpha)(\sigma-\beta)^2 + a_{03}(\sigma-\beta)^3 + a_{40}(\tau-\alpha)^4 + \cdots$$

$$(A2.1.1)$$

Using a similar approach as the one for the single-variable Taylor series, we set τ and σ to α and β and find $f(\alpha, \beta) = a_{00}$. To find the other coefficients we use partial differentiation of f at point α, β:

$$\frac{\partial f(\alpha, \beta)}{\partial \tau} = a_{10}, \quad \frac{\partial f(\alpha, \beta)}{\partial \sigma} = a_{01}, \quad \frac{\partial^2 f(\alpha, \beta)}{\partial \tau^2} = 2a_{20},$$

$$\frac{\partial^2 f(\alpha, \beta)}{\partial \tau \, \partial \sigma} = a_{11}, \quad \frac{\partial^2 f(\alpha, \beta)}{\partial \sigma^2} = 2a_{02}$$

(A2.1.2)

This technique can be used to obtain the full power series of f. In most applications we are interested in the linear approximation of the 2D series:

$$\boxed{f(\tau, \sigma) \approx f(\alpha, \beta) + \frac{\partial f(\alpha, \beta)}{\partial \tau}(\tau - \alpha) + \frac{\partial f(\alpha, \beta)}{\partial \sigma}(\sigma - \beta)}$$

(A2.1.3a)

The higher-order nonlinear terms are often not considered because we assume that we only look at f closely around point α, β; therefore, $\tau - \alpha$ and $\sigma - \beta$ are very small numbers, and higher powers of these small contributions are even smaller. In other words, when f is in the neighborhood of point α, β, the function can be approximated with the linear terms in Equation (A2.13a). In many cases, especially in physics literature, you may encounter an alternative notation for the linear approximation of a nonlinear system. The small fluctuations $\tau - \alpha$ and $\sigma - \beta$ around α, β are indicated as perturbations $\delta\tau$ and $\delta\sigma$, and the notation for $f(\alpha, \beta)$, $(\partial f(\alpha, \beta))/\partial \tau$, and $(\partial f(\alpha, \beta))/\partial \sigma$ is changed to $[f]_{\alpha,\beta}$, $[\partial f/\partial \tau]_{\alpha,\beta}$, and $[\partial f/\partial \sigma]_{\alpha,\beta}$:

$$\boxed{f(\tau, \sigma) \approx [f]_{\alpha,\beta} + \left[\frac{\partial f}{\partial \tau}\right]_{\alpha,\beta} \delta\tau + \left[\frac{\partial f}{\partial \sigma}\right]_{\alpha,\beta} \delta\sigma}$$

(A2.1.3b)

Again, recall that in this notation $[f]_{\alpha,\beta}$, $[\partial f/\partial \tau]_{\alpha,\beta}$, and $[\partial f/\partial \sigma]_{\alpha,\beta}$ represent the coefficients in the equation. They are numbers and not functions, since these represent the function and its derivatives when evaluated at point α, β. An example of an application that linearizes the nonlinear Hodgkin and Huxley equations can be found in Chapter 10 of Koch (1999).

3 Volterra Series

3.1 Introduction

Most physiological systems cannot be modeled successfully as linear systems. At best, a linear model can be considered an approximation of physiological activity in cases where the output of a physiological system behaves (almost) linearly over a limited range of the input. In the following, we extend the convolution integral that describes the behavior of linear devices to the convolution-like Volterra series, which can be used to represent nonlinear systems. Because the expressions for higher-order nonlinear terms require significant computational resources and become very complex to deal with, we will demonstrate the general principles for second-order systems. See Schetzen (2006) if you are interested in details of higher-order systems.

In a linear time invariant (LTI) system, the convolution integral links output $y(t)$ and input $x(t)$ by means of its weighting function $h(t)$ (Fig. 3.1) (Chapter 8 in van Drongelen, 2007):

$$y(t) = h(t) \otimes x(t) = \int\limits_{-\infty}^{\infty} h(\tau)\, x(t - \tau)\mathrm{d}\tau \tag{3.1}$$

Here $\otimes$ symbolizes the convolution operation and the system's weighting function $h(t)$ is its unit impulse response (UIR). This role of $h(t)$ can be verified by using a unit impulse $\delta(t)$ as the input. In this case we obtain (using the sifting property):

$$\int\limits_{-\infty}^{\infty} h(\tau)\delta(t - \tau)\mathrm{d}\tau = h(t) \tag{3.2}$$

Signal Processing for Neuroscientists, A Companion Volume. DOI: 10.1016/B978-0-12-384915-1.00003-6

Note: In the following we will use the sifting property of the unit impulse repeatedly (for a discussion, see section 2.2.2 in van Drongelen, 2007). The sifting property is defined as:

$$x(t) = \int_{-\infty}^{\infty} x(\tau)\delta(\tau - t)d\tau = \int_{-\infty}^{\infty} x(\tau)\delta(t - \tau)d\tau$$

The unit impulse δ has properties of a function with even symmetry; therefore, the evaluation of the integral above is the same for $\delta(t - \tau)$ and $\delta(\tau - t)$. You can also see that this must be the case since the outcome is $\delta(0)$ for $t = \tau$ in both cases, $\delta(t - \tau)$ and $\delta(\tau - t)$.

Such an LTI system shows superposition and scaling properties. For instance, if we introduce a **scaled delta function** $C\delta(t)$ (C—constant) at the input, we get a **scaled UIR function** $Ch(t)$ at the output:

$$\int_{-\infty}^{\infty} h(\tau)C\delta(t - \tau)d\tau = C\int_{-\infty}^{\infty} h(\tau)\delta(t - \tau)d\tau = Ch(t) \tag{3.3}$$

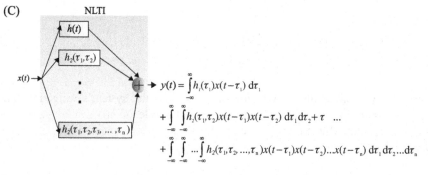

Figure 3.1 Example of LTI and NLTI systems. (A) A linear system. (B) A second-order system and (C) a combined nth order system. The output of the first-order components is determined by the convolution integral, and the output of higher-order components is obtained from convolution-like integrals. The output of the nth order system is represented by a Volterra series consisting of the sum of the individual components, each determined by a convolution-like expression. *(Color in electronic version.)*

Now consider a system that is governed by an equation that is convolution-like:

$$y(t) = \int_{-\infty}^{\infty} \int_{-\infty}^{\infty} h_2(\tau_1, \tau_2) x(t - \tau_1) x(t - \tau_2) d\tau_1 \, d\tau_2 \tag{3.4}$$

Unlike the convolution in Equation (3.1), this system works on two copies of input $x(t)$ instead of only one. As we discussed in Section 2.5, such a system is an example of a so-called second-order Volterra system. Note that the double integral in Equation (3.4) is identical to the last term in the expression in Equation (2.24). If we determine the UIR for the system in Equation (3.4), we get:

$$h_2(t, t) = \int_{-\infty}^{\infty} \int_{-\infty}^{\infty} h_2(\tau_1, \tau_2) \delta(t - \tau_1) \delta(t - \tau_2) d\tau_1 \, d\tau_2 \tag{3.5}$$

Here we applied the sifting property twice: once for each of the delays τ_1 and τ_2. The result $h_2(t,t)$ is the diagonal of kernel h_2.

Note: You can see that in a second-order Volterra system, the UIR $h_2(t,t)$ does not fully characterize the system (unlike the situation in an LTI system). Instead it only characterizes the 2D function $h_2(\tau_1, \tau_2)$ along the diagonal $\tau_1 = \tau_2$ in the τ_1, τ_2 plane. As we will see in Section 3.3 we need sets of paired impulses to fully characterize h_2.

The system in Equation (3.4) is nonlinear because scaling does not hold. For example, the response to a scaled delta function $C\delta(t)$ at the input is:

$$\int_{-\infty}^{\infty} \int_{-\infty}^{\infty} h_2(\tau_1, \tau_2) C\delta(t - \tau_1) C\delta(t - \tau_2) d\tau_1 \, d\tau_2 =$$
$$C^2 \int_{-\infty}^{\infty} \int_{-\infty}^{\infty} h_2(\tau_1, \tau_2) \delta(t - \tau_1) \delta(t - \tau_2) d\tau_1 \, d\tau_2 = C^2 h_2(t, t) \tag{3.6}$$

By comparing Equations (3.5) and (3.6) we can see that in this system the UIR $h_2(t,t)$ scales with C^2 instead of C. As we will show in Section 3.2.1, superposition does not hold for this system either, but showing that scaling does not hold is sufficient to negate linearity of the system. In the remainder of this chapter we will continue our introduction of Section 2.5 by studying the properties of the Volterra series and applying it for the characterization of higher-order systems.

3.2   Volterra Series

The mathematician Vito Volterra used series of convolution-like expressions to define the input−output relationship of NLTI systems:

$$\overbrace{y(t) = \int_{-\infty}^{\infty} h_1(\tau_1)x(t - \tau_1)d\tau_1}^{\text{1st order term}}$$

$$\overbrace{+ \int_{-\infty}^{\infty}\int_{-\infty}^{\infty} h_2(\tau_1,\tau_2)x(t - \tau_1)x(t - \tau_2)d\tau_1\, d\tau_2}^{\text{2nd order term}} + \cdots$$

$$\underbrace{+ \int_{-\infty}^{\infty}\int_{-\infty}^{\infty} \cdots \int_{-\infty}^{\infty} h_n(\tau_1,\tau_2,\ldots,\tau_n)x(t - \tau_1)x(t - \tau_2)\ldots x(t - \tau_n)d\tau_1\, d\tau_2 \ldots d\tau_n}_{n\text{th order term}}$$

$$(3.7)$$

The output $y(t)$ of an nth order system depends on multiple copies of the input and is the sum of the 1st, 2nd, ..., nth order convolution-like expressions. The functions h_1, h_2, ..., h_n are called the 1st, 2nd, ..., nth order **Volterra kernels**. In some texts, a zero-order kernel (h_0) representing a DC term, or offset, is added to Equation (3.7). Just as in an LTI system, $y(t)$ is the UIR if the input $x(t)$ is a unit impulse $\delta(t)$. In higher-order systems, the contribution of the nth order Volterra kernel to the UIR is a so-called diagonal slice through the kernel, that is, a section through the kernel with all delays τ_1, τ_2, ..., τ_n equal. An example for a second-order system ($n = 2$) is shown in Equation (3.5).

Note: We can refer to h_1 as the UIR only if we deal with a first-order Volterra system without a DC term—that is, a (linear) system where h_1 is the only term of $y(t)$. In all other cases, the UIR is determined by the contributions of all of the system's Volterra kernels and not just by h_1.

If we represent the 1st, 2nd, ..., nth order convolution-like terms in Equation (3.7) as H_1, H_2, ..., H_n we get an alternative, simplified notation:

$$y(t) = H_1[x(t)] + H_2[x(t)] + \cdots + H_n[x(t)] \tag{3.8}$$

Equation (3.8) can be generalized for an nth order system:

$$y(t) = \sum_{n=1}^{N} H_n[x(t)] \tag{3.9a}$$

In some cases a DC term $H_0[x(t)] = h_0$ (with h_0 being a constant) is added. This allows one to further generalize Equation (3.8) for the NLTI system to:

$$y(t) = \sum_{n=0}^{N} H_n[x(t)] \tag{3.9b}$$

Just as with any series, we should consider the convergence of the Volterra series. In our case, we approach this by optimistically assuming that any system we consider will be associated with a converging series. We can afford this optimism because we will apply the Volterra series only to known, relatively low-order systems and because we would immediately notice if the output predicted by the Volterra series would poorly match the measured output of the system it represents.

Recall that we can consider the Volterra series' approximation of output as a Taylor series with the addition of memory (Section 2.5). The Taylor series links output with instantaneous input (no memory, Equation (2.16)), whereas the Volterra series includes a memory component in the convolution-like integrals. These integrals show that the output at time t depends not only on current input signal $x(t)$, but on multiple copies of the delayed input, represented by $x(t - \tau_1)$, $x(t - \tau_2), \ldots, x(t - \tau_n)$ in the integrals in Equation (3.7).

3.2.1 Combined Input to a Second-Order Volterra System

In general, the input–output relationship of a second-order Volterra system without lower-order components can be specified by Equation (3.4). We also demonstrated above that a second-order Volterra system does not scale as an LTI system (compare Equations (3.3) and (3.6)). We can next show that the **superposition property** of an LTI system does not hold in the second-order Volterra system either. To accomplish this, we will determine the system's response to the sum of two inputs $x(t) = x_1(t) + x_2(t)$ relative to its responses to $x_1(t)$ and $x_2(t)$ individually. The response to the combined inputs is:

$$
\begin{aligned}
y(t) &= \int_{-\infty}^{\infty} \int_{-\infty}^{\infty} h_2(\tau_1, \tau_2) x(t - \tau_1) x(t - \tau_2) d\tau_1 \, d\tau_2 \\
&= \int_{-\infty}^{\infty} \int_{-\infty}^{\infty} h_2(\tau_1, \tau_2) [x_1(t - \tau_1) + x_2(t - \tau_1)][x_1(t - \tau_2) + x_2(t - \tau_2)] d\tau_1 \, d\tau_2
\end{aligned}
\tag{3.10}
$$

In Equation (3.10) we have the following four terms:

$$H_2[x_1(t)] = \int_{-\infty}^{\infty} \int_{-\infty}^{\infty} h_2(\tau_1, \tau_2) x_1(t - \tau_1) x_1(t - \tau_2) d\tau_1 \, d\tau_2 \tag{3.11a}$$

$$H_2[x_2(t)] = \int\limits_{-\infty}^{\infty} \int\limits_{-\infty}^{\infty} h_2(\tau_1, \tau_2) x_2(t - \tau_1) x_2(t - \tau_2) d\tau_1\, d\tau_2 \qquad (3.11b)$$

$$\left.\begin{array}{l} H_2[x_1(t), x_2(t)] = \displaystyle\int\limits_{-\infty}^{\infty} \int\limits_{-\infty}^{\infty} h_2(\tau_1, \tau_2) x_1(t - \tau_1) x_2(t - \tau_2) d\tau_1\, d\tau_2 \\[2em] H_2[x_2(t), x_1(t)] = \displaystyle\int\limits_{-\infty}^{\infty} \int\limits_{-\infty}^{\infty} h_2(\tau_1, \tau_2) x_1(t - \tau_2) x_2(t - \tau_1) d\tau_1\, d\tau_2 \end{array}\right\} \text{cross-terms}$$

$$(3.11c)$$

Note that Equations (3.11a) and (3.11b) represent the expressions for the system's response when its input would be x_1 and x_2, respectively. The two cross-terms in expression (3.11c) are determined by both x_1 and x_2 and can be considered equal because **the second-order kernel is symmetric**—that is, $h(\tau_1, \tau_2) = h(\tau_2, \tau_1)$.

Note: The symmetry of $h(\tau_1, \tau_2, \ldots, \tau_n)$: Recall that the kernel h of a linear system is obtained from the system's response to a unit impulse. As we will see in the following section, h can be determined in higher-order systems from the responses to multiple unit impulses. Since kernel h can be obtained from responses to combinations of unit impulses, the symmetry assumption makes sense. This is the case because there is no reason to assume that a system would be able to distinguish (i.e., react differently) between unit impulse 1 followed by unit impulse 2 as compared to unit impulse 2 followed by unit impulse 1 (the unit impulses are indistinguishable because they are identical). For a formal explanation see chapter 3 in Schetzen (2006). If you have problems following this reasoning, you may come back to it after studying the concrete example in Pr3_1.m and Section 3.3.

Based on the symmetry, we can rewrite the second equation in (3.11c) as:

$$\int\limits_{-\infty}^{\infty} \int\limits_{-\infty}^{\infty} h_2(\tau_2, \tau_1) x_1(t - \tau_2) x_2(t - \tau_1) d\tau_2\, d\tau_1 \qquad (3.11d)$$

If we now interchange the dummy variables τ_1 and τ_2 this becomes:

$$\int\limits_{-\infty}^{\infty} \int\limits_{-\infty}^{\infty} h_2(\tau_1, \tau_2) x_1(t - \tau_1) x_2(t - \tau_2) d\tau_1\, d\tau_2 \qquad (3.11e)$$

This result indicates that the two expressions in (3.11c) are equal, so we may combine the cross-terms into:

$$2H_2[x_1(t), x_2(t)] = 2 \int\limits_{-\infty}^{\infty} \int\limits_{-\infty}^{\infty} h_2(\tau_1, \tau_2)x_1(t - \tau_1)x_2(t - \tau_2)d\tau_1 \, d\tau_2 \qquad (3.11\text{f})$$

By combining Equations (3.10), (3.11a), (3.11b), and (3.11f), we get the following expression for the output $y(t)$ for the sum of the inputs $x_1(t) + x_2(t)$:

$$y(t) = H_2[x(t)] = H_2[x_1(t) + x_2(t)] = H_2[x_1(t)] + H_2[x_2(t)] + 2H_2[x_1(t), x_2(t)]$$
$$(3.12)$$

The cross-terms $2H_2[x_1(t), x_2(t)]$ in Equation (3.11f) represent the deviation of the second-order Volterra system's response from the response to $x_1(t) + x_2(t)$ if superposition were to hold, that is, in the second-order Volterra system the total response $y(t)$ to $x_1(t) + x_2(t)$ is **not** equal to the sum (superposition) of the responses to $x_1(t)$ and $x_2(t)$ individually: $H_2[x_1(t)] + H_2[x_2(t)]$.

3.3   A Second-Order Volterra System

As we discussed in Chapter 2, we can create a dynamical nonlinear system by combining a dynamical linear system (L) and a static nonlinear (N) one (Figs. 2.2 and 3.2A); the type of system that emerges from this combination is often indicated as an LN cascade. In the example in Fig. 3.2A, the dynamical linear system is a simple low-pass filter consisting of a resistor (R) and capacitor (C) ($h_{RC} = 1/(RC)e^{-t/RC}$). If you need to refresh your basic knowledge about RC filters, see chapters 10 and 11 in van Drongelen (2007). The static nonlinear component in Fig. 3.2A relates an input to the square of the output, that is, output $y(t)$ is the square of its input $f(t) : y(t) = f(t)^2$. From this relationship, the static nonlinear component is considered a squarer. Following the procedure described in Section 2.5, we can establish that this cascade is a second-order Volterra system with a second-order kernel:

$$h_2(\tau_1, \tau_2) = h_{RC}(\tau_1)h_{RC}(\tau_2) = \left(\frac{1}{RC}e^{-\tau_1/RC}\right)\left(\frac{1}{RC}e^{-\tau_2/RC}\right) = \left(\frac{1}{RC}\right)^2 e^{-(\tau_1 + \tau_2)/RC}$$
$$(3.13)$$

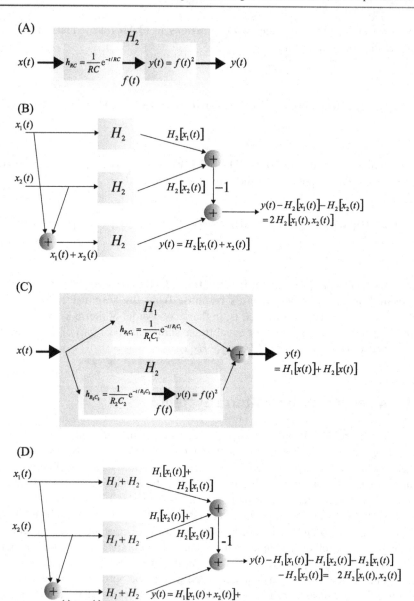

Figure 3.2 (A) Nonlinear system consisting of a cascade of a linear filter (a dynamical system) and a squarer (a static system). (B) Procedure to compute H_2 following Equation (3.12). This procedure is one of the approaches used to determine kernel h_2 in Pr3_1.m. (C) A general second-order Volterra system with a first-order (H_1) and second-order (H_2) operator. (D) The same procedure as in (B) applied to the second-order system with a first-order component. This procedure is followed in script Pr3_2.m. *(Color in electronic version.)*

The following MATLAB code (the first part of Pr3_1.m *available on http:// www.elsevierdirect.com/companions/9780123849151) will create an image of the 2D Volterra kernel (Fig. 3.3) based on the known structure of the nonlinear cascade (Equation (3.13)).*

```
% Linear component is a low-pass RC circuit
% we use R=10k and C=3.3uF
R=10e3;
C=3.3e-6;
RC=R*C;

% Timing parameters
sample_rate=1000;
dt=1/sample_rate;
time=0.1;
A=RC/dt;
T=100;            % The setting for timing and the length of correlation
                  % calculations for both Volterra and Wiener kernels
% Step 1. The analog continuous time approach using the square of
% the unit impulse response of the filter: h(t)=(1/RC)*exp(-t/RC)
% to compare with discrete time approach (in the following Steps) we assume
% the discrete time steps (dt) and interval (time)
j=1;
for tau1=0:dt:time;
    i=1;
    r1(j)=(1/RC)*exp(-tau1/RC);   % 1st-order response in the cascade
    for tau2=0:dt:time
        y(i,j)=((1/RC)*exp(-tau1/RC))*((1/RC)*exp(-tau2/RC));
                  % Output y is h2 (=2nd order Volterra kernel)
                  % which is the square of the filter response
        i=i+1;
    end;
    j=j+1;
end;
% plot the surface of h2
y=y*dt^2;          % scale for the sample value dt
figure; surf(y);
axis([0 T 0 T min(min(y)) max(max(y))])
view(100,50)
title('2nd order Volterra kernel (h2) of an LN cascade')
xlabel('tau1');ylabel('tau2');zlabel('h2');
```

Now we validate our (nonparametric) approach with the Volterra series by using a parametric model, the Wiener cascade, depicted in Fig. 3.2A. So far we assumed that the internal structure of the second-order system is known. In this example we study

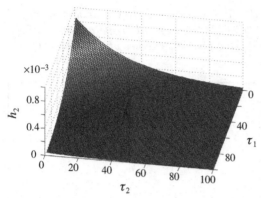

Figure 3.3 Example of a second-order Volterra kernel $h_2(\tau_1, \tau_2)$ determined by Equation (3.13) in MATLAB script Pr3_1.m. *(Color in electronic version.)*

the LN cascade system in Fig. 3.2A but we pretend to know only that it is a second-order Volterra system and that we do **not** know that it is a cascade or what its components are. Now we can use Equation (3.12) and the procedure described in Steps 1–6 below to find the second-order kernel h_2. Finally we will compare this result with the one we can obtain analytically (Equation (3.13)), which is shown in Fig. 3.3.

(1) Using the approach in the example in Section 3.2.1, we use a pair of unit impulses occurring at times T_1 ($\delta(t - T_1)$) and T_2 ($\delta(t - T_2)$) as inputs $x_1(t)$ and $x_2(t)$, respectively.

(2) We determine the system's response to each of the inputs individually: the response to $x_1(t) = \delta(t - T_1)$ is $H_2[\delta(t - T_1)]$ and the response to $x_2(t) = \delta(t - T_2)$ is $H_2[\delta(t - T_2)])$ (Fig. 3.2B).

(3) We determine the response to the sum of both impulses $x_1(t) + x_2(t) = \delta(t - T_1) + \delta(t - T_2)$, which is $y(t) = H_2[\delta(t - T_1) + \delta(t - T_2)]$ and according to Equation (3.12):

$$y(t) = H_2[\delta(t - T_1)] + H_2[\delta(t - T_2)] + 2H_2[\delta(t - T_1), \delta(t - T_2)].$$

(4) From the responses obtained in Steps 2 and 3 above, we can solve for H_2:

$$H_2[\delta(t - T_1), \delta(t - T_2)] = \frac{y(t) - H_2[\delta(t - T_1)] - H_2[\delta(t - T_2)]}{2} \quad (3.14)$$

(5) We use Equation (3.11f) (divided by two) and substitute $\delta(t - T_1)$ for $x_1(t)$ and $\delta(t - T_2)$ for $x_2(t)$:

$$H_2[\delta(t - T_1), \delta(t - T_2)] = \int_{-\infty}^{\infty} \int_{-\infty}^{\infty} h_2(\tau_1, \tau_2)\delta(t - T_1 - \tau_1)\delta(t - T_2 - \tau_2)d\tau_1\,d\tau_2$$

Using the sifting property twice, the double integral evaluates to:

$$\int_{-\infty}^{\infty} \int_{-\infty}^{\infty} h_2(\tau_1, \tau_2)\delta(t - T_1 - \tau_1)\delta(t - T_2 - \tau_2)d\tau_1\,d\tau_2 = h_2(t - T_1, t - T_2) \quad (3.15)$$

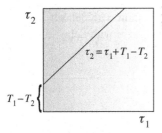

Figure 3.4 The $\tau_1 - \tau_2$ plane and the section represented by $\tau_2 = \tau_1 + T_1 - T_2$. (*Color in electronic version.*)

This is the second-order Volterra kernel we are looking for.

(6) To relate Equations (3.14) and (3.15) to the definition of the second-order kernel $h_2(\tau_1, \tau_2)$, we set $\tau_1 = t - T_1$ and $\tau_2 = t - T_2$. By using the common variable t, we can relate the delays by: $\tau_1 + T_1 = \tau_2 + T_2 \rightarrow \tau_2 = \tau_1 + T_1 - T_2$. In the $\tau_1 - \tau_2$ plane, this represents a line at 45° with an intercept at $T_1 - T_2$.

Therefore, the response obtained in Equation (3.15) is a slice of the second-order kernel along the line $\tau_2 = \tau_1 + T_1 - T_2$.

Following this procedure, we can obtain the second-order kernel by repeatedly probing the system with pairs of unit impulses at different times T_1 and T_2. By varying the timing of the impulses, we can determine h_2 in both dimensions τ_1 and τ_2, that is, we fill in the plane in Fig. 3.4.

3.3.1 Discrete Time Implementation

Now we are ready to implement the procedure described in Steps 1−6 above for the cascade in Fig. 3.2A. A diagram of this procedure is shown in Fig. 3.2B. Because in this example we know the parameters of the LN cascade, we can compare the result we obtain following Steps 1−6 with the earlier analytically derived result based on our knowledge of the system (depicted in Fig. 3.3).

Recall that for the discrete time solution in the MATLAB file below it is assumed that the sample interval dt is much smaller than the time constant of the filter, that is, $RC/\mathrm{d}t \gg 1$ (see section 11.2.2 in van Drongelen, 2007). If this assumption is violated too much, the approximation of the differential equation by the difference equation will be compromised.

Equation (3.14) can be used to determine the second-order kernel of the system. The following MATLAB code (the second part in `Pr3_1.m` *available on http://www.elsevierdirect.com/companions/9780123849151) will create an image of the 2D kernel shown in Fig. 3.5.*

```
i=1; j=0;
delay1=1;

for delay2=delay1:1:length(x);
    j=j+1;
```

```
        x1=zeros(1,100);x1(delay1)=1;     % unit impulse train with delay 1
        x2=zeros(1,100);x2(delay2)=1;     % unit impulse train with delay 2
% The summed input xs, containing two unit impulses
        if (delay1==delay2);
            xs=zeros(1,100);xs(delay1)=2;  % delays are equal
        else
            xs=zeros(1,100);xs(delay1)=1;xs(delay2)=1;
                                    % sum of two unit impulses if delays
                                    % are NOT equal
        end;
% Compute the system outputs to individual and combined unit impulses
        y1_previous=0;
        y2_previous=0;
        ys_previous=0;

        for n=1:length(x);
            % response to delay1
            y1(n)=(A*y1_previous+x1(n))/(A+1);        % the linear component
            y1_previous=y1(n);
            z1(n)=y1(n)^2;                            % the squarer
            % response to delay2
            y2(n)=(A*y2_previous+x2(n))/(A+1);        % the linear component
            y2_previous=y2(n);
            z2(n)=y2(n)^2;                            % the squarer
            % response to the sum of both delays
            ys(n)=(A*ys_previous+xs(n))/(A+1);        % the linear component
            ys_previous=ys(n);
            zs(n)=ys(n)^2;                            % the squarer
        end;

        h=(zs-z1-z2)/2;                    % A slice of the kernel h2
                                           % in the tau1-tau2 plane this is a line
                                           % at 45 degrees with intersection
                                           % delay1-delay2

        tau1=delay2:1:length(x);
        tau2=tau1+(delay1-delay2);
        h=h(delay2:length(h));

        plot3(tau1,tau2,h);
end;

axis([0 T 0 T])
view(100,50)
% Only half is shown because kernel h2 is symmetric
```

```
title('half of 2nd order Volterra kernel (h2) of an LN cascade')
xlabel('tau1');ylabel('tau2');zlabel('h2');
grid on
```

3.4   General Second-Order System

The example of the cascade in Fig. 3.2A has a second-order operator only. Generally a second-order system consists of both a first- and second-order operator (assuming again that there is no H_0 component). Following the notation in Equation (3.9a) with $N = 2$ we get:

$$y(t) = H_1[x(t)] + H_2[x(t)] \tag{3.16}$$

An example of such a system where the H_2 operator (representing an LN cascade) is extended with a first-order component H_1 is shown in Fig. 3.2C.

3.4.1   Determining the Second-Order Kernel

For determining h_2 in a system such as that described by Equation (3.16), we can still use the procedure discussed in Steps 1−6 (Section 3.3) and depicted in Fig. 3.2D. The method still works because superposition holds for the contribution of the first-order operator—that is, for input $x(t) = x_1(t) + x_2(t)$, the contribution of

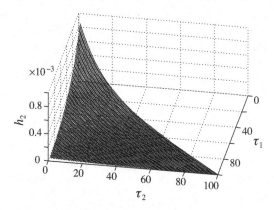

Figure 3.5 By following the procedures in the first and second parts in script Pr3_1.m we can compare the second-order Volterra kernels obtained from the parametric LN cascade model (Fig. 3.3, based on Equation (3.13)) and the one obtained using the nonparametric approach in which we determined h_2 by the procedure outlined in Steps 1−6 (Section 3.3) and represented in Fig. 3.2B. The result of the latter procedure (obtained in the second part of Pr3_1) is shown here. As can be seen by comparing this result with the earlier one in Fig. 3.3, both approaches agree. Because of the symmetry in h_2, only half of the kernel is depicted. *(Color in electronic version.)*

the first-order operator is simply the sum of the contributions for $x_1(t)$ and $x_2(t)$ separately:

$$\int_{-\infty}^{\infty} h_1(\tau)x(t-\tau)\mathrm{d}\tau = \int_{-\infty}^{\infty} h_1(\tau)[x_1(t-\tau)+x_2(t-\tau)]\mathrm{d}\tau$$

$$= \int_{-\infty}^{\infty} h_1(\tau)x_1(t-\tau)\mathrm{d}\tau + \int_{-\infty}^{\infty} h_1(\tau)x_2(t-\tau)\mathrm{d}\tau \qquad (3.17\mathrm{a})$$

or in a more compact notation:

$$H_1[x(t)] = H_1[x_1(t)+x_2(t)] = H_1[x_1(t)]+H_2[x_2(t)] \qquad (3.17\mathrm{b})$$

If we apply the same procedure (as shown in Fig. 3.2B) to a system that obeys $y(t) = H_1[x(t)] + H_2[x(t)]$ (e.g., the system in Fig. 3.2C), the contribution of the first-order kernel will cancel because of the superposition property (Equations (3.17a) and (3.17b)). Just as in the previous example, the output will be $2H_2[x_1(t), x_2(t)]$, allowing us to find the second-order kernel by dividing the output of the procedure by 2 (see Equation (3.14)). In program Pr3_2.m (available on http://www.elsevierdirect.com/companions/9780123849151), the procedure depicted in Fig. 3.2D is followed for the second-order system shown in Fig. 3.2C.

3.4.2 Determining the First-Order Kernel

After we determined the system's second-order kernel, we can also find its first-order kernel via the system's UIR. The UIR of the system in Fig. 3.2C will consist of a first- and second-order component (Fig. 3.6). Therefore, if we determine the system's UIR and subtract its second-order component, we have the first-order Volterra kernel h_1. The second-order component of the system's UIR is the slice through h_2 for

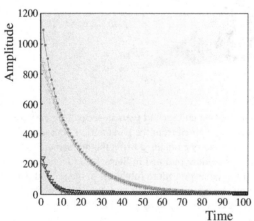

Figure 3.6 An example of a UIR (red dots, upper curve) of a second-order system such as in Fig. 3.2C. The response consists of a first-order component (black triangles, lower curve) and a second-order part (green triangles, middle curve). This result was obtained with MATLAB script Pr3_2.m, albeit with different parameters than the version available on http://www.elsevierdirect.com/companions/9780123849151. (*Color in electronic version.*)

$\tau_1 = \tau_2$ (i.e., the diagonal of the second-order kernel). This approach is now feasible, since we determined h_2 in the previous procedure. To summarize, we find h_1 by:

$$h_1 = \text{UIR} - h_2(\tau_1, \tau_2) \quad \text{for } \tau_1 = \tau_2 \tag{3.18}$$

3.5 System Tests for Internal Structure

Nonlinear systems are usually complex, and to facilitate their characterization, one may attempt to simplify their structure by presenting it as a cascade of basic modules. As we discussed in this chapter and in Section 2.5, we often represent dynamical nonlinear systems with cascades of dynamical linear systems and static nonlinearities. In neuroscience, such cascades are frequently used to model neurons and their networks. For example, the integrate-and-fire neuronal model (e.g., Izhikevich, 2007) combines a linear low-pass filter (RC circuit) to mimic subthreshold integration of the biological membrane combined with a static nonlinearity that generates an action potential when the membrane potential exceeds a threshold. Models for neuronal networks also frequently use the cascade approach. For example, in a model to explain the EEG's alpha rhythm, Lopes da Silva et al. (1974) model synaptic function in the thalamo-cortical network with linear filters and a static nonlinearity to model action potential generation (see their fig. 7). Examples of systems that are frequently used to represent nonlinear systems are depicted in Fig. 3.7; in this section we will discuss how these basic configurations may be recognized by examination of their Volterra kernels.

3.5.1 The LN Cascade

The LN cascade (linear system followed by a nonlinear system, Fig. 3.7A), also called a Wiener system (not to be confused with the Wiener series we will discuss in Chapter 4), was also used in Section 2.5 when we demonstrated that the system's input−output relationship fits the Volterra series representation (Equation (2.24)). This result is repeated here:

$$
\begin{aligned}
z(t) &= a_0 + \int_{-\infty}^{\infty} a_1 h(\tau)x(t - \tau)\mathrm{d}\tau + \int_{-\infty}^{\infty}\int_{-\infty}^{\infty} a_2 h(\tau_1)h(\tau_2)x(t - \tau_1)x(t - \tau_2)\mathrm{d}\tau_1\,\mathrm{d}\tau_2 \\
&= h_0 + \int_{-\infty}^{\infty} h_1(\tau)x(t - \tau)\mathrm{d}\tau + \int_{-\infty}^{\infty}\int_{-\infty}^{\infty} h_2(\tau_1, \tau_2)x(t - \tau_1)x(t - \tau_2)\mathrm{d}\tau_1\,\mathrm{d}\tau_2
\end{aligned}
$$

$$\tag{3.19}$$

From Equation (3.19) we can see that the second-order Volterra kernel h_2 is related to the first-order kernel h_1. The first-order kernel is proportional with the UIR of the Wiener system's linear component: $h_1(\tau) = a_1 h(\tau)$, while the second-order kernel is given by $h_2(\tau_1, \tau_2) = a_2 h(\tau_1)h(\tau_2)$. If we keep one of the variables τ_1 or τ_2 constant, we obtain a section (slice) through the second-order kernel, which is also

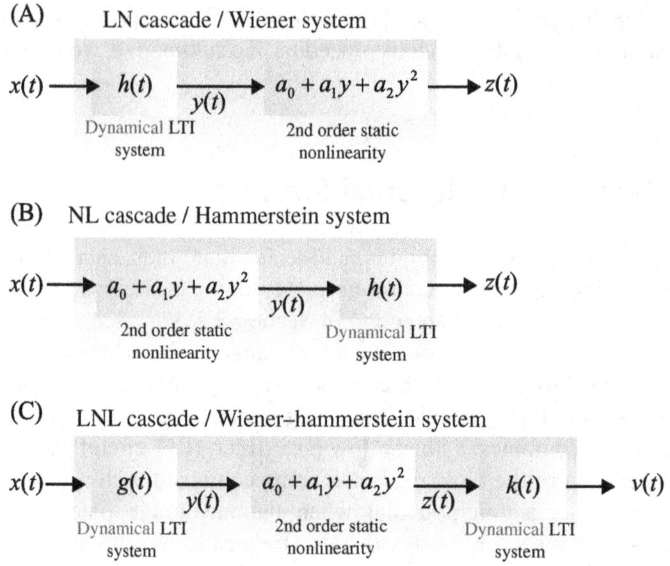

(A)   LN cascade / Wiener system

$$x(t) \longrightarrow \boxed{h(t)} \xrightarrow[y(t)]{} \boxed{a_0 + a_1 y + a_2 y^2} \longrightarrow z(t)$$

Dynamical LTI   2nd order static
system   nonlinearity

(B)   NL cascade / Hammerstein system

$$x(t) \longrightarrow \boxed{a_0 + a_1 y + a_2 y^2} \xrightarrow[y(t)]{} \boxed{h(t)} \longrightarrow z(t)$$

2nd order static   Dynamical LTI
nonlinearity   system

(C)   LNL cascade / Wiener–hammerstein system

$$x(t) \longrightarrow \boxed{g(t)} \xrightarrow[y(t)]{} \boxed{a_0 + a_1 y + a_2 y^2} \xrightarrow[z(t)]{} \boxed{k(t)} \longrightarrow v(t)$$

Dynamical LTI   2nd order static   Dynamical LTI
system   nonlinearity   system

Figure 3.7 Frequently used cascade models to analyze nonlinear dynamical systems. (A) Cascade of a dynamical linear system followed by a static nonlinearity. (B) A similar cascade, but compared with (A) the order has changed: first the static nonlinearity followed by the linear component. (C) A static nonlinearity sandwiched in between two dynamical linear systems. *(Color in electronic version.)*

proportional with the linear component's UIR h. Let us keep τ_2 constant so that $h(\tau_2)$ is a constant value b; we then obtain the expression for a slice through the second-order kernel parallel to the τ_1 axis: $h_2(\tau_1) = ba_2 h(\tau_1)$. It is straightforward to show that the ratio between the first-order kernel and a slice (parallel to the τ_1 axis) of the second-order kernel is the constant a_1/ba_2. It is important to note here that this constant may be negative or positive; this should be taken into account when looking for proportionality. A similar result can be obtained for a slice parallel to the τ_2 axis when we hold τ_1 constant. It should be noted that this condition must be satisfied for a Wiener system but there are other configurations that may show the same property. Therefore, strictly speaking, the condition of proportionality of first-order kernels and second-order slices can be used only to exclude the Wiener structure of a nonlinear system. Optimistically, one might say that if the condition is satisfied for a particular nonlinear system, we may use the Wiener structure to model the system.

3.5.2   The NL Cascade

The cascade shown in Fig. 3.7B, also called a Hammerstein system, is a cascade of a nonlinear static component followed by a linear dynamic one. The output y of the first (nonlinear) component becomes the input of the linear dynamical system. The output from this final dynamical component is then the system's output z:

$$z(t) = \int_{-\infty}^{\infty} h(\tau)y(t-\tau)d\tau = \int_{-\infty}^{\infty} h(\tau)[a_0 + a_1x(t-\tau) + a_2x(t-\tau)^2]d\tau \quad (3.20)$$

If we separate the three terms in Equation (3.20), we can identify the three Volterra kernels h_0, h_1, and h_2.

$$\text{First term:} \quad \underbrace{\int_{-\infty}^{\infty} h(\tau)a_0 \, d\tau}_{h_0} \quad (3.21a)$$

$$\text{Second term:} \quad \int_{-\infty}^{\infty} \underbrace{h(\tau)a_1}_{h_1(\tau)} x(t-\tau) \, d\tau \quad (3.21b)$$

$$\text{Third term:} \quad \int_{-\infty}^{\infty} h(\tau)a_2x(t-\tau)^2 \, d\tau$$

$$= \int_{-\infty}^{\infty} \int_{-\infty}^{\infty} \underbrace{h(\tau_1)a_2\delta(\tau_1 - \tau_2)}_{h_2(\tau_1,\tau_2)} x(t-\tau_1)x(t-\tau_2)d\tau_1 \, d\tau_2$$

$$(3.21c)$$

To obtain the Volterra formalism, we rewrote the single integral expression in Equation (3.21c) as a double integral by separating the product $x(t-\tau)^2$ into $x(t-\tau_1)x(t-\tau_2)$. To make sure this product is only nonzero for $\tau_1 = \tau_2$, we added the $\delta(\tau_1 - \tau_2)$ function. The diagonal slice in the $\tau_1 - \tau_2$ plane of the Hammerstein's second-order kernel ($h(\tau)a_2$) is proportional to the UIR of the cascade's linear component ($h(\tau)$). It can be seen in Equation (3.21b) that the first-order Volterra kernel ($h(\tau)a_1$) is also proportional to the linear component's impulse response. Consequently, the diagonal slice of the second-order kernel is proportional with the first-order kernel. Both characteristics discussed above (nonzero second-order kernel along the diagonal and its proportionality with the first-order kernel) may be used to test an unknown nonlinear system for an underlying Hammerstein structure.

3.5.3 The LNL Cascade

A combination of both the cascades discussed above is shown in Fig. 3.7C. Such a system is an LNL cascade, also called a Wiener–Hammerstein model. We obtain the output $z(t)$ of the static nonlinearity inside the cascade by following the same procedure we used to determine the Wiener system's output (see Equations (2.24) and (3.19)):

$$z(t) = a_0 + a_1 y(t) + a_2 y(t)^2$$

$$= a_0 + a_1 \int_{-\infty}^{\infty} g(\tau)x(t - \tau)d\tau + a_2 \int_{-\infty}^{\infty} \int_{-\infty}^{\infty} g(\tau_1)g(\tau_2)x(t - \tau_1)x(t - \tau_2)d\tau_1\, d\tau_2$$

$$(3.22)$$

The LNL cascade's final output v is then the convolution of the expression above with the UIR k of the second linear system $v(t) = \int_{-\infty}^{\infty} k(\lambda)z(t - \lambda)d\lambda$ (we use λ here for the delay). Using Equation (3.22) for $z(t - \lambda)$ gives:

$$v(t) = a_0 \int_{-\infty}^{\infty} k(\lambda)d\lambda + a_1 \int_{-\infty}^{\infty} \int_{-\infty}^{\infty} k(\lambda)g(\tau)x(t - \tau - \lambda)d\tau\, d\lambda \ldots$$

$$(3.23)$$

$$+ a_2 \int_{-\infty}^{\infty} \int_{-\infty}^{\infty} \int_{-\infty}^{\infty} k(\lambda)g(\tau_1)g(\tau_2)x(t - \tau_1 - \lambda)x(t - \tau_2 - \lambda)d\tau_1\, d\tau_2\, d\lambda$$

To simplify, the first-order part of this expression can be rewritten using $\omega = \lambda + \tau$:

$$a_1 \int_{-\infty}^{\infty} \int_{-\infty}^{\infty} k(\lambda)g(\omega - \lambda)x(t - \omega)d\omega\, d\lambda = \int_{-\infty}^{\infty} \underbrace{\left[a_1 \int_{-\infty}^{\infty} k(\lambda)g(\omega - \lambda)d\lambda \right]}_{h_1(\omega)} x(t - \omega)d\omega$$

$$(3.24a)$$

Similarly, using $v = \lambda + \tau_1$ and $\omega = \lambda + \tau_2$, the second-order part becomes:

$$a_2 \int_{-\infty}^{\infty} \int_{-\infty}^{\infty} \int_{-\infty}^{\infty} k(\lambda)g(v - \lambda)g(\omega - \lambda)x(t - v)x(t - \omega)dv\, d\omega\, d\lambda$$

$$= \int_{-\infty}^{\infty} \int_{-\infty}^{\infty} \underbrace{\left[a_2 \int_{-\infty}^{\infty} k(\lambda)g(v - \lambda)g(\omega - \lambda)d\lambda \right]}_{h_2(v,\omega)} x(t - v)x(t - \omega)dv\, d\omega$$

$$(3.24b)$$

We can see that the second-order kernel is the integral expression in between the brackets: $a_2 \int_{-\infty}^{\infty} k(\lambda)g(v - \lambda)g(\omega - \lambda)d\lambda$. From this expression we can obtain the so-called second-order marginal kernel K_2^m (the sum of all kernel slices over one of the variables). If we integrate this expression with respect to one of its variables, say v, we get:

$$K_2^m = a_2 \int_{-\infty}^{\infty} \int_{-\infty}^{\infty} k(\lambda)g(v-\lambda)g(\omega-\lambda)d\lambda\,dv \tag{3.25a}$$

Now we make a change of timing variables again, $\xi = v - \lambda$, $d\xi = dv$, and rearrange the integral operation:

$$a_2 \int_{-\infty}^{\infty} \int_{-\infty}^{\infty} k(\lambda)g(v-\lambda)g(\omega-\lambda)d\lambda\,dv = a_2 \int_{-\infty}^{\infty} k(\lambda) \underbrace{\left[\int_{-\infty}^{\infty} g(\xi)d\xi\right]}_{A} g(\omega-\lambda)d\lambda$$

$$= a_2 \underbrace{A}_{\text{I}} \underbrace{\left[\int_{-\infty}^{\infty} k(\lambda)g(\omega-\lambda)d\lambda\right]}_{\text{II}} \tag{3.25b}$$

In the first step, we regrouped the integral operations and defined the outcome of the integral with respect to $d\xi$ as a number A. Subsequently we separated the expression into two parts. Part I is equal to A and Part II is proportional with the expression for the first-order kernel; this relationship can be seen by comparing Part II with the expression for h_1 in Equation (3.24a): $a_1 \int_{-\infty}^{\infty} k(\lambda)g(\omega - \lambda)d\lambda$. This latter term is simply Part II scaled by a_1. Of course, we would have obtained a similar outcome had we integrated the second-order kernel with respect to ω. This reasoning leads us to conclude that in an LNL sandwich, the marginal kernel K_2^m (the summation [integral] of all slices of the second-order kernel h_2 parallel to one of the axes) is proportional with the first-order kernel h_1. We can use the above proportionality between the marginal second-order kernel and the first-order one to test for a potential underlying sandwich structure of an unknown system. Because other types of cascade may show a similar property, this will allow us to exclude an LNL sandwich structure or to make it likely that we are dealing with one.

3.6 Sinusoidal Signals

When we use a sinusoidal signal as the input to a linear system, we get a sinusoidal signal at its output. At the output, the amplitude of the sinusoidal signal may be amplified/attenuated and the waveform may have a changed phase, but the frequencies of the input and output of an LTI system are identical. We can use this property to completely characterize an LTI system, such as the RC filter, with a set of sinusoidal inputs (see section 10.3 in van Drongelen, 2007). Since the frequency at the output does not change relative to the input frequency, we can describe the LTI system by depicting change of amplitude and phase for each frequency with a Bode plot or an Nyquist plot (see section 12.3, fig. 12.4 in van Drongelen, 2007).

As you may have guessed, this simple relationship between input and output frequency is not valid for nonlinear systems. Let us investigate the response of the second-order nonlinear system introduced in Equation (3.5) by feeding it a cosine with amplitude A and angular frequency ω_0. Further, let us use Euler's relationship $(e^{\pm j\phi} = \cos\phi \pm j\sin\phi)$ to express the input in terms of two complex exponentials:

$$x(t) = A\cos\omega_0 t = \underbrace{\frac{A}{2}e^{j\omega_0 t}}_{x_1(t)} + \underbrace{\frac{A}{2}e^{-j\omega_0 t}}_{x_2(t)} \tag{3.26}$$

Note that the two components of input x (x_1 and x_2) are complex conjugates. Now we can treat this input as we did in Section 3.2.1 and repeat the result from Equation (3.10) for the system's output y:

$$y(t) = \int_{-\infty}^{\infty}\int_{-\infty}^{\infty} h_2(\tau_1, \tau_2)[x_1(t-\tau_1)+x_2(t-\tau_1)][x_1(t-\tau_2)+x_2(t-\tau_2)]d\tau_1\,d\tau_2$$

$$\tag{3.27}$$

In short notation we can write:

$$y(t) = H_2[x_1(t)] + H_2[x_2(t)] + H_2[x_1(t), x_2(t)] + H_2[x_2(t), x_1(t)] \tag{3.28}$$

The only difference between Equation (3.28) and the Equation (3.12) obtained in Section 3.2.1 is that we did not use the symmetry property $H_2[x_1(t), x_2(t)] = H_2[x_2(t), x_1(t)]$. Let us then evaluate each of the four terms in Equation (3.28). The first term is:

$$\begin{aligned}H_2[x_1(t)] &= \int_{-\infty}^{\infty}\int_{-\infty}^{\infty} h_2(\tau_1,\tau_2)x_1(t-\tau_1)x_1(t-\tau_2)d\tau_1\,d\tau_2\\ &= \left(\frac{A}{2}\right)^2 \int_{-\infty}^{\infty}\int_{-\infty}^{\infty} h_2(\tau_1,\tau_2)e^{j\omega_0(t-\tau_1)}\,e^{j\omega_0(t-\tau_2)}\,d\tau_1\,d\tau_2\end{aligned} \tag{3.29}$$

Combining both exponential expressions we get:

$$\left(\frac{A}{2}\right)^2 e^{j2\omega_0 t} \underbrace{\int_{-\infty}^{\infty}\int_{-\infty}^{\infty} h_2(\tau_1,\tau_2)e^{-j\omega_0\tau_1}\,e^{-j\omega_0\tau_2}\,d\tau_1\,d\tau_2}_{\Psi} = \left(\frac{A}{2}\right)^2 e^{j2\omega_0 t}\,\Psi(-j\omega_0, -j\omega_0)$$

$$\tag{3.30a}$$

Here we use the variable Ψ to symbolize the double integral, a complex function of ω_0.

Note: Comparing the function Ψ above (symbolizing the double integral) with equation (6.4) in van Drongelen (2007), it can be seen that the expression is the 2D Fourier transform of the second-order kernel.

Similarly, substituting the exponential expression for x_2, the second term $H_2[x_2(t)]$ in Equation (3.28) becomes:

$$\left(\frac{A}{2}\right)^2 e^{-j2\omega_0 t} \int\limits_{-\infty}^{\infty} \int\limits_{-\infty}^{\infty} h_2(\tau_1, \tau_2) e^{j\omega_0 \tau_1} e^{j\omega_0 \tau_2} \, d\tau_1 \, d\tau_2 = \left(\frac{A}{2}\right)^2 e^{-j2\omega_0 t} \Psi(j\omega_0, j\omega_0)$$

(3.30b)

Note that both Equations (3.30a) and (3.30b) include an exponent in which the frequency is doubled ($2\omega_0$ instead of ω_0).

The third term in Equation (3.28) is:

$$
\begin{aligned}
H_2[x_1(t), x_2(t)] &= \int\limits_{-\infty}^{\infty} \int\limits_{-\infty}^{\infty} h_2(\tau_1, \tau_2) x_1(t - \tau_1) x_2(t - \tau_2) d\tau_1 \, d\tau_2 \\
&= \left(\frac{A}{2}\right)^2 \int\limits_{-\infty}^{\infty} \int\limits_{-\infty}^{\infty} h_2(\tau_1, \tau_2) e^{j\omega_0(t - \tau_1)} e^{-j\omega_0(t - \tau_2)} \, d\tau_1 \, d\tau_2
\end{aligned}
$$

(3.31)

Combining the exponentials in the expression above, we get:

$$\left(\frac{A}{2}\right)^2 \int\limits_{-\infty}^{\infty} \int\limits_{-\infty}^{\infty} h_2(\tau_1, \tau_2) e^{-j\omega_0 \tau_1} e^{j\omega_0 \tau_2} \, d\tau_1 \, d\tau_2 = \left(\frac{A}{2}\right)^2 \Psi(-j\omega_0, j\omega_0) \qquad (3.32a)$$

Using the same approach the fourth term becomes:

$$\left(\frac{A}{2}\right)^2 \int\limits_{-\infty}^{\infty} \int\limits_{-\infty}^{\infty} h_2(\tau_1, \tau_2) e^{j\omega_0 \tau_1} e^{-j\omega_0 \tau_2} \, d\tau_1 \, d\tau_2 = \left(\frac{A}{2}\right)^2 \Psi(j\omega_0, -j\omega_0) \qquad (3.32b)$$

Substituting the results for all four terms obtained in (3.30a), (3.30b), (3.32a), and (3.32b) into Equation (3.28) we now have:

$$
\begin{aligned}
y(t) = & \left[\left(\frac{A}{2}\right)^2 e^{j2\omega_0 t} \Psi(-j\omega_0, -j\omega_0) + \left(\frac{A}{2}\right)^2 e^{-j2\omega_0 t} \Psi(j\omega_0, j\omega_0) \right] \\
& + \left[\left(\frac{A}{2}\right)^2 \Psi(-j\omega_0, j\omega_0) + \left(\frac{A}{2}\right)^2 \Psi(j\omega_0, -j\omega_0) \right]
\end{aligned}
$$

(3.33)

It can be seen that the first two terms and the second two terms (grouped by brackets) are the complex conjugates of each other. Therefore, we may conclude that the expression in Equation (3.33) is real since the sum of two complex conjugates is real (the sum of imaginary numbers $a + jb$ and $a - jb$ is $2a$). Consequently we get the following result:

$$y(t) = 2\left(\frac{A}{2}\right)^2 Re(e^{j2\omega_0 t}\Psi(-j\omega_0, -j\omega_0)) + 2\left(\frac{A}{2}\right)^2 Re(\Psi(-j\omega_0, j\omega_0)) \qquad (3.34)$$

in which $Re(\ldots)$ denotes the real component. Using Euler's relationship again, we can see that the output contains a sinusoid:

$$y(t) = 2\left(\frac{A}{2}\right)^2 Re[(\cos 2\omega_0 t + j\sin 2\omega_0 t)\Psi(-j\omega_0, -j\omega_0)] + 2\left(\frac{A}{2}\right)^2 Re[\Psi(-j\omega_0, j\omega_0)]$$
$$(3.35)$$

The output contains a constant (the second term in Equation (3.35)) and a sinusoid with a frequency $2\omega_0$ (the first term).

The expression in Equation (3.35) is an important result for the analysis of higher-order systems: a certain frequency at the system's input may result in a higher-frequency component at the output. **When we digitize the output of a higher-order system as the result of some input signal, it is important to estimate the highest frequency at the output to avoid aliasing** (Section 2.2.2 in van Drongelen, 2007). With a linear system, this problem does not occur; the highest frequency of the input is the highest frequency possible at the output. But with nonlinear systems, the maximum output frequency may be a multiple (as shown above, in a second-order system it is a factor of two, and in an nth order system it is a factor of n) of the input's highest frequency value. A practical approach here is to first sample the system's output at a much higher sample rate than would be used routinely (one to a few orders of magnitude higher) and then compute a power spectrum to estimate the highest frequency component. The outcome of this preliminary experiment can be used to establish an appropriate sample rate.

4 Wiener Series

4.1 Introduction

Determining the Volterra kernels of an unknown system faces several practical problems: (1) the order of the system underlying the signal being investigated is usually unknown, and (2) the contributions of the individual components (of different order) of the Volterra series are not independent. The first problem is generally an issue if one wants to characterize a system with any type of series approximation, and the second problem can sometimes be resolved by the use of a series with orthogonal components. An example of the latter procedure is the development of the Fourier series; by having orthogonal terms, there are no dependencies between terms and one can determine the coefficients a_i and b_i of the Fourier series sequentially (Chapter 5, van Drongelen, 2007). To address the dependence between components in a series approximation for nonlinear systems, **Norbert Wiener developed an approach where each component in his Volterra-like series is orthogonal to all lower-order ones.** Although within the Wiener series approach one cannot predetermine the order of the system being studied either (problem (1) above), the orthogonality between the terms in the series allows one to determine the Wiener kernels sequentially. Subsequently one can determine their contribution to the signal, and stop the kernel-estimation process at the order where the signal is sufficiently approximated. For practical reasons most studies limit their kernel estimates at either the second order or (less often) at the third order. The third- and higher-order kernels require significant computation and they are difficult to depict.

In this chapter we will first discuss the main differences between the Wiener and Volterra series, after which we will derive the expressions for the zero-, first-, and second-order Wiener kernels, and then finally we will discuss practical methods for determining Wiener kernels for simulated and recorded time series. Applications of these methods will be presented in the form of MATLAB scripts. The last part of this chapter and Fig. 4.6 summarize the mathematical procedures we use to determine the Wiener series. For an extended background on this topic, see Marmarelis and Marmarelis (1978) and the reprint of Schetzen's book (Schetzen, 2006), and for recent engineering-oriented overviews see Westwick and Kearney (2003) and Marmarelis (2004).

Signal Processing for Neuroscientists, A Companion Volume. DOI: 10.1016/B978-0-12-384915-1.00004-8

4.2   Wiener Kernels

Similar to the Volterra series, the Wiener series characterizes the output z of a nonlinear system as the sum of a set of operators G_n dependent on kernels k_n and input x:

$$z(t) = \sum_{n=0}^{N} G_n[k_n; x(t)]$$

This equation is similar to the ones for the Volterra series (Equations (3.9a) and (3.9b)), but although there are many similarities between Volterra and Wiener series, there are also a few crucial differences that allow Wiener operators to be mutually independent. For clarity we first summarize the **three major differences** between the Volterra and Wiener series and then explain further details in the remainder of this chapter (e.g., the exact relationship between Volterra and Wiener kernels is discussed in Section 4.5).

The first major difference is that although a Volterra series usually does not include a zero-order term (see Equation (3.9)), we may define such a term as a constant:

$$H_0[x(t)] = h_0 \qquad (4.1a)$$

In contrast, the Wiener series **always** includes a zero-order term. This term is defined as the average output (the DC component) equal to k_0:

$$\boxed{G_0[k_0; x(t)] = k_0} \qquad (4.1b)$$

In this equation, k_0 is the zero-order Wiener kernel. We use k_n for the Wiener kernels to distinguish them from the Volterra kernels h_n.

The second major difference is that while individual Volterra operators are homogeneous (see, e.g., Equation (3.6) for a second-order one) (i.e., $H_n[cx(t)] = c^n H_n[x(t)]$), the Wiener operators are nonhomogeneous—for example, the first-order Wiener operator has a first-order and a derived zero-order component:

$$G_1[k_1; x(t)] = g_1[k_1, k_{0(1)}; x(t)] = K_1[x(t)] + K_{0(1)}[x(t)]$$

$$= \int_{-\infty}^{\infty} k_1(\tau_1) x(t - \tau_1) \mathrm{d}\tau_1 + k_{0(1)} \qquad (4.2)$$

The subscript $0(1)$ in $K_{0(1)}$ and $k_{0(1)}$ indicates that these are zero-order members of a first-order nonhomogeneous operator. Specifically, k_1 is the first-order Wiener kernel and $k_{0(1)}$ is the so-called derived Wiener kernel from operator G_1. In general, the Wiener kernels of the type $k_{n(m)}$ with $n < m$ are called derived Wiener kernels because, as we will see below, they must be derived from Wiener kernel k_m. The notation with the capital G indicates that the operator includes both the kernel and input, while the lower-case notation g differs by explicitly also indicating all of the derived kernels.

The second-order Wiener operator is:

$$
\begin{aligned}
G_2[k_2; x(t)] = g_2[k_2, k_{1(2)}, k_{0(2)}; x(t)] &= K_2[x(t)] + K_{1(2)}[x(t)] + K_{0(2)}[x(t)] \\
&= \int\limits_{-\infty}^{\infty} \int\limits_{-\infty}^{\infty} k_2(\tau_1, \tau_2) x(t - \tau_1) x(t - \tau_2) \mathrm{d}\tau_1 \mathrm{d}\tau_2 \\
&+ \int\limits_{-\infty}^{\infty} k_{1(2)}(\tau_1) x(t - \tau_1) \mathrm{d}\tau_1 + k_{0(2)}
\end{aligned}
\tag{4.3}
$$

The subscripts 0(2) and 1(2) indicate that these are zero- and first-order members (derived Wiener kernels) of the second-order nonhomogeneous operator G_2, respectively. Kernel k_2 is the second-order Wiener kernel, while $k_{1(2)}$ and $k_{0(2)}$ are derived Wiener kernels from operator G_2. As we will demonstrate below, the rationale for using nonhomogeneous operators relates to their orthogonality.

The third and final major difference is that in the case of a Wiener series we use a special input signal, usually in the form of zero mean Gaussian white noise (GWN) (alternative input signals are discussed in Section 4.7 and Chapter 5). Selection of a special input is critical because it allows us to create a series in which the **operators are orthogonal (uncorrelated) to the lower-order operators**. As we will see in Section 4.3, this property contributes to creating independence between the operators in the series, which will allow us to determine the Wiener kernels sequentially without having to worry about dependency issues.

The first-order Wiener operator is defined so that it is orthogonal to the zero-order Volterra operator:

$$
E\{H_0[x(t)]g_1[k_1, k_{0(1)}; x(t)]\} = \langle H_0[x(t)]g_1[k_1, k_{0(1)}; x(t)] \rangle = 0
\tag{4.4a}
$$

In the expression after the equal sign, $\langle \ldots \rangle$ indicates the time average.

Note: $\langle x(t) \rangle$ represents the time average of a signal $x(t)$ over a time interval T. This is an alternative notation for the integral notation: $(1/T) \int\limits_0^T x(t) \mathrm{d}t$.

Equation (4.4a) indicates that we assumed ergodicity so that we may use a time average $\langle H_0[x(t)]g_1[k_1, k_{0(1)}; x(t)] \rangle$ to determine the Expectation $E\{\ldots\}$ of the product of H_0 and g_1. If you need to review the concepts of Expectation and time averages, see section 3.2 and appendix 3.1 in van Drongelen (2007). Details about time averages for GWN are reviewed in Appendix 4.1.

Similarly, the second-order Wiener operator is defined as orthogonal to zero- and first-order Volterra operators:

$$\langle H_0[x(t)]g_2[k_2, k_{1(2)}, k_{0(2)}; x(t)]\rangle = 0 \tag{4.4b}$$

$$\langle H_1[x(t)]g_2[k_2, k_{1(2)}, k_{0(2)}; x(t)]\rangle = 0 \tag{4.4c}$$

To characterize any nonlinear system of order N, this approach is generalized for all Wiener operators; that is, for zero mean GWN input, operator $G_n[k_n; x(t)] = g_n[k_n, k_{n-1(n)}, \ldots, k_{0(n)}; x(t)]$ is defined such that it is orthogonal to **any** Volterra operator of a lower order:

$$\langle H_m[x(t)]g_n[k_n, k_{n-1(n)}, \ldots, k_{0(n)}; x(t)]\rangle = 0 \quad \text{for } m < n \tag{4.4d}$$

In the following sections we will achieve orthogonality between the G operators and lower-order Volterra operators by using the so-called Gram−Schmidt technique (for details of this technique, see, e.g., Arfken and Weber, 2005). By defining the Wiener kernels according to this technique, we can determine the kernels of nonlinear systems from lower to higher order without knowledge of the system's components. This procedure is similar to approximating a function or signal with a Fourier series (van Drongelen, 2007, Chapter 5) or a polynomial (Section 2.4). For each kernel (for each order) we can determine its contribution to the output and we can continue to add higher-order terms until we are satisfied with our approximation of the system at hand. The procedure for determining the Wiener kernels as sketched above and their independence to lower-order kernels is summarized in Fig. 4.1. In the following sections we derive the expressions for the first- and second-order Wiener kernels. If you are interested in higher-order components, see Schetzen (2006).

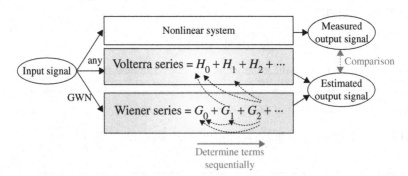

Figure 4.1 Representation of a nonlinear system by Volterra and Wiener series. In contrast to the Volterra operators H_n, the operators G_n in the Wiener series are independent from lower-order operators (stippled arrows). This allows one to determine the Wiener operators and their kernels sequentially and compute their contribution to the estimated output. The comparison with the measured output signal can be used to determine at what order the output is sufficiently approximated. (*Color in electronic version.*)

4.2.1   Derivation of the First-Order Wiener Operator

In the previous section we already identified the zero-order kernel as the signal's DC component in Equation (4.1b). Now we can use orthogonality defined in Equation (4.4) to derive the Wiener kernels k_1 and k_2. Starting with the first-order kernel, we substitute Equations (4.1a) and (4.2) in Equation (4.4a) and find that the following condition must be satisfied:

$$\langle H_0[x(t)]g_1[k_1,k_{0(1)};x(t)]\rangle = 0 \rightarrow$$

$$\left\langle h_0\left[\int_{-\infty}^{\infty} k_1(\tau_1)x(t-\tau_1)d\tau_1 + k_{0(1)}\right]\right\rangle = h_0\left[\int_{-\infty}^{\infty} k_1(\tau_1)\langle x(t-\tau_1)\rangle d\tau_1 + k_{0(1)}\right] = 0$$

$$(4.5)$$

Note that in the expression after the equal sign we took all constants ($h_0, k_1(\tau_1)$, $k_{0(1)}$) out of the time average operation, such that only the (time-dependent) input time series x remains within the time average brackets $\langle\ldots\rangle$. **Now you will see how convenient it is to have zero mean GWN as input.** Because input x is zero mean GWN, the time average $\langle x(t-\tau_1)\rangle$ is zero. Therefore the integral in Equation (4.5) evaluates to zero and (since h_0 is generally not zero) we find that the orthogonality condition in Equation (4.5) is satisfied when:

$$\boxed{k_{0(1)} = 0}$$
$$(4.6)$$

Combining this result with Equation (4.2), we find that the first-order Wiener operator is:

$$\boxed{G_1[k_1;x(t)] = g_1[k_1;x(t)] = \int_{-\infty}^{\infty} k_1(\tau_1)x(t-\tau_1)d\tau_1}$$
$$(4.7)$$

Note that for a first-order system (without a DC component) k_1 is the UIR (van Drongelen, 2007, Chapter 8). Further, if the input x is zero mean GWN, the output of the first-order (linear) operator G_1 will also be zero mean GWN. Because $\langle x(t-\tau_1)\rangle = 0$, the Expectation or the time average of G_1 is zero: that is, $E\{G_1\} = \langle G_1\rangle = 0$. Therefore G_1 is indeed orthogonal to any constant, such as zero-order operators G_0 and H_0.

4.2.2   Derivation of the Second-Order Wiener Operator

We can obtain the expression for the second-order operator g_2 using a procedure similar to the one we developed for the first-order one. Here we must deal separately with the independence between the second-order Wiener operator and the two lower-order Volterra operators H_0 and H_1, respectively.

4.2.2.1 Orthogonality Between H_0 and g_2

Using Equations (4.1a), (4.3), and (4.4b) we get:

$$\langle H_0[x(t)]g_2[k_2, k_{1(2)}, k_{0(2)}; x(t)]\rangle$$

$$= \left\langle h_0 \left[\int_{-\infty}^{\infty} \int_{-\infty}^{\infty} k_2(\tau_1, \tau_2)x(t - \tau_1)x(t - \tau_2)d\tau_1\, d\tau_2 \right.\right.$$

$$\left.\left. + \int_{-\infty}^{\infty} k_{1(2)}(\tau_1)x(t - \tau_1)d\tau_1 + k_{0(2)} \right] \right\rangle = 0$$

$$= h_0 \left[\int_{-\infty}^{\infty} \int_{-\infty}^{\infty} k_2(\tau_1, \tau_2)\langle x(t - \tau_1)x(t - \tau_2)\rangle d\tau_1\, d\tau_2 \right.$$

$$\left. + \int_{-\infty}^{\infty} k_{1(2)}(\tau_1)\langle x(t - \tau_1)\rangle d\tau_1 + k_{0(2)} \right] = 0 \qquad (4.8)$$

As we did in Equation (4.5), we took the constants out of the time average $\langle \ldots \rangle$ such that only the time series x remains within it. Because the input is zero mean GWN with variance σ^2, the average $\langle x(t - \tau_1)\rangle = 0$ and the averaged product of both copies of input x is the autocorrelation R_{xx} (see van Drongelen, 2007, section 8.4.1):

$$\langle x(t - \tau_1)x(t - \tau_2)\rangle = R_{xx}(\tau_2 - \tau_1) = \sigma^2\delta(\tau_2 - \tau_1).$$

Again we can see how convenient the zero mean GWN input is: the time average $\langle x(t - \tau_1)\rangle$ vanishes and time average $\langle x(t - \tau_1)x(t - \tau_2)\rangle$ can be simplified to the expression for the autocorrelation. (See also Appendix 4.1 for further details on averages of products of Gaussian variables.) Therefore Equation (4.8) becomes:

$$h_0 \int_{-\infty}^{\infty} \int_{-\infty}^{\infty} k_2(\tau_1, \tau_2) \underbrace{\langle x(t - \tau_1)x(t - \tau_2)\rangle}_{R_{xx}(\tau_2 - \tau_1)} d\tau_1\, d\tau_2 + h_0k_{0(2)}$$

$$= \sigma^2 h_0 \int_{-\infty}^{\infty} \int_{-\infty}^{\infty} k_2(\tau_1, \tau_2)\delta(\tau_2 - \tau_1)d\tau_1\, d\tau_2 + h_0k_{0(2)}$$

The double integral on the right-hand side can be evaluated by using the sifting property while evaluating the integral for one of the time constants; here we integrate with respect to τ_2 and get:

$$\sigma^2 h_0 \int_{-\infty}^{\infty} k_2(\tau_1, \tau_1)d\tau_1 + h_0k_{0(2)} = 0 \rightarrow \boxed{k_{0(2)} = -\sigma^2 \int_{-\infty}^{\infty} k_2(\tau_1, \tau_1)d\tau_1} \qquad (4.9)$$

In this expression we can see that $k_{0(2)}$ is indeed a *derived* Wiener kernel because it is directly derived from Wiener kernel k_2.

4.2.2.2 Orthogonality Between H_1 and g_2

Subsequently we substitute expression for the first-order Volterra operator (see Equation (3.7)) and Equation (4.3) for the second-order Wiener operator in the orthogonality condition in Equation (4.4c):

$$\langle H_1[x(t)]g_2[k_2, k_{1(2)}, k_{0(2)}; x(t)]\rangle$$

$$= \left\langle \left[\int_{-\infty}^{\infty} h_1(v)x(t-v)dv\right]\left[\int_{-\infty}^{\infty}\int_{-\infty}^{\infty} k_2(\tau_1,\tau_2)x(t-\tau_1)x(t-\tau_2)d\tau_1\,d\tau_2\right.\right.$$

$$\left.\left.+ \int_{-\infty}^{\infty} k_{1(2)}(\tau_1)x(t-\tau_1)d\tau_1 + k_{0(2)}\right]\right\rangle \tag{4.10}$$

The above expression contains three terms. We will first show that the first and third terms always evaluate to zero if the input is zero mean GWN.

The **first** term:

$$\left\langle \left[\int_{-\infty}^{\infty} h_1(v)x(t-v)dv\right]\left[\int_{-\infty}^{\infty}\int_{-\infty}^{\infty} k_2(\tau_1,\tau_2)x(t-\tau_1)x(t-\tau_2)d\tau_1\,d\tau_2\right]\right\rangle \tag{4.11a}$$

$$= \int_{-\infty}^{\infty}\int_{-\infty}^{\infty}\int_{-\infty}^{\infty} h_1(v)k_2(\tau_1,\tau_2)\langle x(t-v)x(t-\tau_1)x(t-\tau_2)\rangle dv\,d\tau_1\,d\tau_2 = 0$$

evaluates to zero because of our choice of zero mean GWN as input and the odd product $\langle x(t-v)x(t-\tau_1)x(t-\tau_2)\rangle = 0$ (Appendix 4.1)—**again taking advantage of our choice of GWN as the input**.

The **third** term in Equation (4.10):

$$\left[\int_{-\infty}^{\infty} h_1(v)\langle x(t-v)\rangle dv\right]k_{0(2)} = 0 \tag{4.11b}$$

also evaluates to zero because $\langle x(t-v)\rangle = 0$.

The **second** term in Equation (4.10) is:

$$\left\langle \left[\int_{-\infty}^{\infty} h_1(v)x(t-v)dv\right]\left[\int_{-\infty}^{\infty} k_{1(2)}(\tau_1)x(t-\tau_1)d\tau_1\right]\right\rangle \tag{4.11c}$$

$$= \int_{-\infty}^{\infty}\int_{-\infty}^{\infty} h_1(v)k_{1(2)}(\tau_1)\langle x(t-v)x(t-\tau_1)\rangle dv\,d\tau_1$$

This second term is the only one that contains an even product of $x(t)$ and can be further evaluated using (again) the autocorrelation R_{xx} for the zero mean GWN with variance σ^2; that is, $\langle x(t - v)x(t - \tau_1) \rangle = R_{xx}(\tau_1 - v) = \sigma^2 \delta(\tau_1 - v)$. This gives us:

$$\sigma^2 \int_{-\infty}^{\infty} \int_{-\infty}^{\infty} h_1(v)k_{1(2)}(\tau_1)\delta(\tau_1 - v)dv \, d\tau_1 = \sigma^2 \int_{-\infty}^{\infty} h_1(\tau_1)k_{1(2)}(\tau_1)d\tau_1$$

In the above we evaluate the integral with respect to v by using the sifting property. Because the first and third terms already evaluate to zero, the second term must be zero in order to satisfy the orthogonality condition in Equation (4.4c). We accomplish this by setting:

$$\boxed{k_{1(2)} = 0} \tag{4.12}$$

Substituting the results we obtained from the orthogonality conditions (in Equations (4.9) and (4.12)) into Equation (4.3), we find the second-order Wiener operator G_2:

$$\begin{array}{c} G_2[k_2; x(t)] = g_2[k_2, k_{1(2)}, k_{0(2)}; x(t)] \\[2mm] = \int_{-\infty}^{\infty} \int_{-\infty}^{\infty} k_2(\tau_1, \tau_2)x(t - \tau_1)x(t - \tau_2)d\tau_1 \, d\tau_2 - \underbrace{\sigma^2 \int_{-\infty}^{\infty} k_2(\tau_1, \tau_1)d\tau_1}_{k_{0(2)}} \end{array} \tag{4.13}$$

Note that just as the Expectation for G_1 is zero, the expected output of G_2 is also zero:

$$E\left\{ \int_{-\infty}^{\infty} \int_{-\infty}^{\infty} k_2(\tau_1, \tau_2)x(t - \tau_1)x(t - \tau_2)d\tau_1 \, d\tau_2 - \sigma^2 \int_{-\infty}^{\infty} k_2(\tau_1, \tau_1)d\tau_1 \right\}$$

$$= \int_{-\infty}^{\infty} \int_{-\infty}^{\infty} k_2(\tau_1, \tau_2)E\{x(t - \tau_1)x(t - \tau_2)\}d\tau_1 \, d\tau_2 - \sigma^2 \int_{-\infty}^{\infty} k_2(\tau_1, \tau_1)d\tau_1 \tag{4.14}$$

As for the time average of an even product of Gaussian variables, once again the Expectation is the autocorrelation: $E\{x(t - \tau_1)x(t - \tau_2)\} = \sigma^2 \delta(\tau_1 - \tau_2)$. Substituting this into Equation (4.14) gives:

$$\int_{-\infty}^{\infty} \int_{-\infty}^{\infty} k_2(\tau_1, \tau_2)\sigma^2 \delta(\tau_1 - \tau_2)d\tau_1 \, d\tau_2 - \sigma^2 \int_{-\infty}^{\infty} k_2(\tau_1, \tau_1)d\tau_1$$

$$= \sigma^2 \int_{-\infty}^{\infty} k_2(\tau_1, \tau_1)d\tau_1 - \sigma^2 \int_{-\infty}^{\infty} k_2(\tau_1, \tau_1)d\tau_1 = 0 \tag{4.15}$$

Here we evaluated the term with the double integral using the sifting property. Because the Expectation is zero, G_2 (just as G_1) is orthogonal to any constant including G_0. Even without knowledge about our derivation above, using the same approach, it is straightforward to show that operator G_2 is also designed to be orthogonal to G_1. This orthogonality can be evaluated via the Expectation of the product $E\{G_1, G_2\}$, which contains odd products of the random input variable $x(t)$. **The odd products evaluate to zero (Appendix 4.1), causing the Expectation to vanish.**

In the above we showed that the first-order Wiener operator is orthogonal to the zero-order one, and that the second-order operator is orthogonal to the first- and zero-order ones. We will not elaborate on this here, but in general the Wiener operators are constructed in a way that they are orthogonal to all lower-order ones. Higher-order Wiener kernels will not be derived here, but the derivation follows a similar procedure as described for the zero- to second-order kernels above. Details for these derivations can be found in Schetzen (2006).

4.3   Determination of the Zero-, First- and Second-Order Wiener Kernels

Now that we know how the expressions for the terms in the Wiener series are developed, it is time to examine how we might determine the terms from measured and simulated data sets. Recall also that we can determine the kernels sequentially because of the orthogonality property. The best-known method to establish Wiener kernels from measurements is the cross-correlation method first described by Lee and Schetzen (1965). If we deal with a nonlinear system of order N, and we present a zero mean GWN x at its input, we obtain output z as the sum of the Wiener operators G_n:

$$z(t) = \sum_{n=0}^{N} G_n[k_n; x(t)] \tag{4.16}$$

4.3.1   Determination of the Zero-Order Wiener Kernel

As we extend our results for the first- and second-order operators to all higher-order ones, we find that the Expectation of all Wiener operators G_n, except the zero-order operator G_0, is zero (see the last paragraph in Sections 4.2.1 and 4.2.2.2). Therefore, assuming an ergodic process (allowing the use of time averages for estimating Expectations), we find that the average of output signal z is:

$$\langle z(t) \rangle = \sum_{n=0}^{N} \langle G_n[k_n; x(t)] \rangle = G_0[k_0; x(t)] = k_0 \tag{4.17}$$

Thus the zero-order Wiener kernel is obtained from the mean output (i.e., the output's DC component).

4.3.2 Determination of the First-Order Wiener Kernel

Here we show that we can get the first-order Wiener kernel of a system from the cross-correlation between its input x and output z:

$$
\begin{aligned}
\langle z(t)x(t - v_1)\rangle &= \langle G_0[k_0; x(t)]x(t - v_1)\rangle + \langle G_1[k_1; x(t)]x(t - v_1)\rangle \\
&\quad + \langle G_2[k_2; x(t)]x(t - v_1)\rangle + \cdots \\
&= \sum_{n=0}^{N}\langle G_n[k_n; x(t)]x(t - v_1)\rangle
\end{aligned}
\tag{4.18}
$$

Recall that Wiener kernels are defined to be orthogonal to lower-order Volterra kernels. This property may be generalized to **all** lower-order Volterra kernels. Since the delay operator $x(t - v_1)$ can be presented as a first-order Volterra operator (Appendix 4.2), all Wiener operators G_n with $n \geq 2$ are orthogonal to $x(t - v_1)$ according to Equation (4.4d). Let us check this property by examining the outcome for $\langle z(t)x(t - v_1)\rangle$ by determining the outcome for the individual operators $G_0, G_1, G_2, \ldots$

Using Equation (4.1b) for $n = 0$:

$$
\langle G_0[k_0; x(t)]x(t - v_1)\rangle = k_0 \underbrace{\langle x(t - v_1)\rangle}_{0} = 0
\tag{4.19a}
$$

Using Equation (4.7) for $n = 1$:

$$
\begin{aligned}
\langle G_1[k_1; x(t)]x(t - v_1)\rangle &= \int_{-\infty}^{\infty} k_1(\tau_1)\underbrace{\langle x(t - \tau_1)x(t - v_1)\rangle}_{\sigma^2\delta(\tau_1 - v_1)}\,\mathrm{d}\tau_1 \\
&= \sigma^2 \int_{-\infty}^{\infty} k_1(\tau_1)\delta(\tau_1 - v_1)\mathrm{d}\tau_1 = \sigma^2 k_1(v_1)
\end{aligned}
\tag{4.19b}
$$

For the second-order kernel we already know that $\langle G_1[k_1; x(t)]x(t - v_1)\rangle$ is zero because $x(t - v_1)$ can be considered a lower-order Volterra operator (Appendix 4.2). However, let us check the outcome of the cross-correlation anyway.

Using Equation (4.13) for $n = 2$:

$$
\begin{aligned}
&\langle G_2[k_2; x(t)]x(t - v_1)\rangle \\
&= \int_{-\infty}^{\infty}\int_{-\infty}^{\infty} k_2(\tau_1, \tau_2)\underbrace{\langle x(t - \tau_1)x(t - \tau_2)x(t - v_1)\rangle}_{0}\,\mathrm{d}\tau_1\,\mathrm{d}\tau_2 \\
&\quad - \sigma^2 \int_{-\infty}^{\infty} k_2(\tau_1, \tau_1)\underbrace{\langle x(t - v_1)\rangle}_{0}\,\mathrm{d}\tau_1 = 0
\end{aligned}
\tag{4.19c}
$$

The above integrals evaluate to zero because they contain time average of odd products of the GWN input x. We now state (without further checking) that the remaining averaged products ($\langle G_n[k_n; x(t)]x(t - v_1)\rangle$, $n \geq 3$) are zero by using the property of Equation (4.4d). From the three expressions in Equations (4.19a) and (4.19c) we conclude that only the term for $n = 1$ is nonzero; therefore, we can determine the first-order Wiener kernel by combining Equations (4.18) and (4.19b):

$$\langle z(t)x(t - v_1)\rangle = \sigma^2 k_1(v_1) \rightarrow \boxed{k_1(v_1) = \frac{1}{\sigma^2}\langle z(t)x(t - v_1)\rangle} \qquad (4.20)$$

Thus the first-order Wiener kernel can be found by the cross-correlation between input x and output z weighted by the variance of the input.

4.3.3 Determination of the Second-Order Wiener Kernel

Using an analogous procedure for the higher-order Wiener kernels, we can find the second-order kernel by using a (second-order) cross-correlation between output z and now two copies of input x:

$$\langle z(t)x(t - v_1)x(t - v_2)\rangle$$
$$= \langle G_0[k_0; x(t)]x(t - v_1)x(t - v_2)\rangle + \langle G_1[k_1; x(t)]x(t - v_1)x(t - v_2)\rangle$$
$$+ \langle G_2[k_2; x(t)]x(t - v_1)x(t - v_2)\rangle + \cdots = \sum_{n=0}^{N}\langle G_n[k_n; x(t)]x(t - v_1)x(t - v_2)\rangle$$

$$(4.21)$$

Because $x(t - v_1)x(t - v_2)$ can be presented as a second-order Volterra operator (Appendix 4.2), all Wiener operators G_n with $n \geq 3$ are orthogonal to $x(t - v_1)x(t - v_2)$ according to Equation (4.4d).

Using Equation (4.1b) for $n = 0$:

$$\langle G_0[k_0; x(t)]x(t - v_1)x(t - v_2)\rangle = k_0 \underbrace{\langle x(t - v_1)x(t - v_2)\rangle}_{\sigma^2\delta(v_1 - v_2)} = k_0\sigma^2\delta(v_1 - v_2)$$

$$(4.22a)$$

Using Equation (4.7) for $n = 1$:

$$\langle G_1[k_1; x(t)]x(t - v_1)x(t - v_2)\rangle = \int_{-\infty}^{\infty} k_1(\tau_1)\underbrace{\langle x(t - \tau_1)x(t - v_1)x(t - v_2)\rangle}_{0}\, d\tau_1 = 0$$

$$(4.22b)$$

Using Equation (4.13) for $n = 2$:

$$\langle G_2[k_2; x(t)]x(t - v_1)x(t - v_2)\rangle$$

$$= \underbrace{\int_{-\infty}^{\infty} \int_{-\infty}^{\infty} k_2(\tau_1, \tau_2) \underbrace{\langle x(t - \tau_1)x(t - \tau_2)x(t - v_1)x(t - v_2)\rangle}_{A} \, d\tau_1 \, d\tau_2}_{I}$$

$$- \sigma^2 \underbrace{\int_{-\infty}^{\infty} k_2(\tau_1, \tau_1) \underbrace{\langle x(t - v_1)x(t - v_2)\rangle}_{\sigma^2\delta(v_1 - v_2)} \, d\tau_1}_{II} \qquad (4.22c)$$

Using Wick's theorem (a theorem that relates higher-order moments to lower-order ones; Appendix 4.1), the average indicated by A in Equation (4.22c) (fourth-order correlation) can be written as:

$$A = \langle x(t - \tau_1)x(t - \tau_2)x(t - v_1)x(t - v_2)\rangle$$

$$= \underbrace{\langle x(t - \tau_1)x(t - \tau_2)\rangle}_{\sigma^2\delta(\tau_1 - \tau_2)} \underbrace{\langle x(t - v_1)x(t - v_2)\rangle}_{\sigma^2\delta(v_1 - v_2)}$$

$$+ \underbrace{\langle x(t - \tau_1)x(t - v_1)\rangle}_{\sigma^2\delta(\tau_1 - v_1)} \underbrace{\langle x(t - \tau_2)x(t - v_2)\rangle}_{\sigma^2\delta(\tau_2 - v_2)}$$

$$+ \underbrace{\langle x(t - \tau_1)x(t - v_2)\rangle}_{\sigma^2\delta(\tau_1 - v_2)} \underbrace{\langle x(t - \tau_2)x(t - v_1)\rangle}_{\sigma^2\delta(\tau_2 - v_1)}$$

This allows us to separate Part I of the expression in Equation (4.22c) into the following three terms:

(1) $\sigma^4 \int_{-\infty}^{\infty} \int_{-\infty}^{\infty} k_2(\tau_1, \tau_2)\delta(\tau_1 - \tau_2)\delta(v_1 - v_2)d\tau_1 \, d\tau_2 = \sigma^4 \int_{-\infty}^{\infty} k_2(\tau_1, \tau_1)\delta(v_1 - v_2)d\tau_1$

(2) $\sigma^4 \int_{-\infty}^{\infty} \int_{-\infty}^{\infty} k_2(\tau_1, \tau_2)\delta(\tau_1 - v_1)\delta(\tau_2 - v_2)d\tau_1 \, d\tau_2 = \sigma^4 k_2(v_1, v_2)$

(3) $\sigma^4 \int_{-\infty}^{\infty} \int_{-\infty}^{\infty} k_2(\tau_1, \tau_2)\delta(\tau_1 - v_2)\delta(\tau_2 - v_1)d\tau_1 \, d\tau_2 = \sigma^4 k_2(v_2, v_1) = \sigma^4 k_2(v_1, v_2)$

The three integrals above are evaluated using the sifting property. Furthermore, by using the same symmetry property of the Volterra kernels (Section 3.2.1), we have concluded that k_2 is symmetrical and that the terms in 2 and 3 above are identical.

Combining the results for $1-3$ in Part I and the integral term Part II in Equation (4.22c) we get:

$$\langle G_2[k_2; x(t)]x(t - v_1)x(t - v_2)\rangle$$

$$= \underbrace{2\sigma^4 k_2(v_1, v_2) + \sigma^4 \int_{-\infty}^{\infty} k_2(\tau_1, \tau_1)\delta(v_1 - v_2)d\tau_1}_{\text{I}}$$

$$\underbrace{- \sigma^2 \int_{-\infty}^{\infty} k_2(\tau_1, \tau_1) \underbrace{\langle x(t - v_1)x(t - v_2)\rangle}_{\sigma^2\delta(v_1 - v_2)} d\tau_1}_{\text{II}} \qquad (4.22d)$$

The integral terms in the expression above cancel so that the final result becomes:

$$\langle G_2[k_2; x(t)]x(t - v_1)x(t - v_2)\rangle = 2\sigma^4 k_2(v_1, v_2) \qquad (4.22e)$$

According to Equation (4.4d) all Wiener operators for $n > 2$ are defined so that their contributions will be zero. This allows us to combine Equations (4.21) with (4.22a), (4.22b), and (4.22e):

$$\langle z(t)x(t - v_1)x(t - v_2)\rangle = k_0\sigma^2\delta(v_1 - v_2) + 2\sigma^4 k_2(v_1, v_2) \qquad (4.23)$$

Now we decide to ignore the case when $v_1 = v_2$ and assume that $v_1 \neq v_2$ so that $\delta(v_1 - v_2) = 0$. In this case the first term on the right-hand side of Equation (4.23) evaluates to zero. Therefore, for the off-diagonal part ($v_1 \neq v_2$) of the second-order Wiener kernel we have:

$$\boxed{k_2(v_1, v_2) = \frac{1}{2\sigma^4} \langle z(t)x(t - v_1)x(t - v_2)\rangle \quad \text{for } v_1 \neq v_2} \qquad (4.24)$$

The second-order Wiener kernel is the second-order cross-correlation between output and input weighted by $2\sigma^4$. Our trick to ignore the diagonally located terms may seem a bit strange but in practical applications, **the limitation imposed by $v_1 \neq v_2$ does not present a problem because we can compute k_2 for delays that are arbitrarily close to $v_1 = v_2$.**

4.4 Implementation of the Cross-Correlation Method

In this section we present a practical application for finding Wiener kernels associated with a given nonlinear system (Fig. 4.2). MATLAB implementations of this approach are in Pr4_1.m and Pr4_2.m. In principle we can use Equations (4.17), (4.20), and (4.24) to determine the Wiener kernels. However, since our input of random noise is necessarily finite, the subsequent kernels may not be exactly orthogonal. To mitigate the effects of this problem, it is common practice to

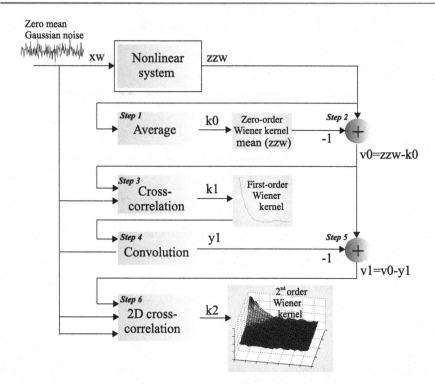

Figure 4.2 Lee−Schetzen cross-correlation method for obtaining the zero-, first-, second-order Wiener kernels of a nonlinear system. Zero mean GWN is used as the input (xw) of a nonlinear system. The average of the output (zzw) is used to estimate the zero-order kernel k0 . The residue v0 (= zzw-k0) is then cross-correlated with the input to estimate the first-order kernel k1 . Subsequently, the contribution y1 of the first-order kernel is determined by convolving it with the input. Finally, the residue v1 (= v0-y1) is correlated with two copies of the input (2D cross-correlation) for the estimation of k2 . (*Color in electronic version.*)

determine the kernels from low to higher orders sequentially while at each step subtracting the contribution of the lower-order kernels from the output (-1, $+$ operations in Fig. 4.2). For example, before computing k_1, it is common practice **to subtract k_0** (the DC component) from the output to obtain a zero-order residue v_0. This residue v_0 ($= z - k_0$), instead of the output z, is then cross-correlated with the input to obtain the first-order kernel k_1 (recall Equation (4.20)):

$$k_1(v_1) = \frac{1}{\sigma^2} \langle v_0(t)x(t - v_1) \rangle \qquad (4.25)$$

To estimate the first-order kernel's contribution (y_1) to the output, the first-order kernel k_1 is convolved with the input x: $y_1 = x \otimes k_1$. This first-order contribution y_1 is then subtracted from the zero-order residue v_0 to obtain the first-order residue v_1.

The residue v_1 ($= z - k_0 - y_1$) is now cross-correlated with two copies of the input to estimate the second-order kernel k_2:

$$k_2(v_1, v_2) = \frac{1}{2\sigma^4} \langle v_1(t)x(t - v_1)x(t - v_2) \rangle \qquad (4.26)$$

Note that as in Equation (4.24), the above expression is valid only for off-diagonal values with $\boxed{v_1 \neq v_2.}$ This procedure is followed in the MATLAB programs and is depicted in Fig. 4.2. The input is variable xw; the output is zzw. The zero- and first-order residues are v0 and v1, respectively. The Wiener kernels are k0, k1, and k2.

An example of a MATLAB implementation can be found in Pr4_1.m and Pr4_2.m. A snippet of the latter is shown here.

```
%%%%%%%%%%%%%%%%%%%%%%%%%%%%%%%%%%%%%%%
%%%%%%%%%%%%% Estimation of the Wiener kernel estimation using
%%%%%%%%%%%%%% the Lee, Schetzen cross-correlation method
%%%%%%%%%%%%%%%%%%%%%%%%%%%%%%%%%%%%%%%%%%%%%%
% First create a set of input output using random noise
xw=randn(10000,1);      % create array with Gaussian white noise
xw=xw-mean(xw);
N=length(xw);
st=std(xw);

figure;subplot(2,1,1),plot(xcorr(xw),'k');
title('Autocorrelation of the Input Shows a Random Noise Characteristic');
subplot(2,1,2);hist(xw);
title('Amplitude Distribution of the Input -> Gaussian');

yw_previous1=0;
yw_previous2=0;
for n=1:length(xw);
   ywh1(n)=(A1*yw_previous1+xw(n))/(A1+1);      % the 1st order operator
   yw_previous1=ywh1(n);
   ywh2(n)=(A2*yw_previous2+xw(n))/(A2+1);      % the linear component of
                                                % the 2nd order operator
   yw_previous2=ywh2(n);
   zzw(n)=ywh1(n)+ywh2(n)^2;      % 1st order component+the squarer
end;

figure; hold;
plot(xw,'k');plot(zzw,'r')
title('Input (black) and Output (red) of a Wiener System')
xlabel('Time (ms)');ylabel('Amplitude')
```

```
%%%%%% The Lee Schetzen Cross-correlation Method
%————————————————————————————————————————
% Step 1 (Fig. 4.1): Determine 0-order Wiener kernel
%————————————————————————————————————————
k0=mean(zzw)
y0=ones(1,length(xw))*k0;
% Step 2 (Fig. 4.1): Subtract k0 from the response to find residue % v0——
%————————————————————————————————————————
v0=zzw-k0;
% Step 3 (Fig. 4.1): Estimate k1 by first-order
% cross-correlation of v0 and input
%————————————————————————————————————————
for i=0:T-1
  temp=0;
  for n=i+1:N
    temp=temp+v0(n)*xw(n-i);
  end;
  k1(i+1)=temp/(N*st^2);
end;

figure; plot(k1);
title('first-order Wiener kernel')

% Step 4 (Fig. 4.1): Compute the output of the first-order
%                Wiener kernel using convolution
%————————————————————————————————————————
for n=1:N;
  temp=0;
  for i=0:min([n-1 T-1]);
    temp=temp+k1(i+1)*xw(n-i);
  end;
  y1(n)=temp;
end;

% Step 5 (Fig. 4.1): Compute the first-order residue
%————————————————————————————————————————
v1=v0-y1;
% Step 6 (Fig. 4.1): Estimate k2 by second-order cross-correlation
%                of v1 with the input
%————————————————————————————————————————
for i=0:T-1
  for j=0:i
    temp=0;
    for n=i+1:N
```

```
        temp=temp+v1(n)*xw(n-i)*xw(n-j);
    end;
    k2(i+1,j+1)=temp/(2*N*st^4);
    k2(j+1,i+1) = k2(i+1,j+1);
  end;
end;

figure; surf(k2(1:T,1:T));
title('second-order Wiener Kernel');
view(100,50);
```

The MATLAB script Pr4_2.m computes the Wiener kernels for a combined system such as the cascade discussed in the previous chapter (Fig. 3.2C). In this example we use low-pass filters for the linear components and a squarer for the nonlinear one (Fig. 4.3).

In the example in Fig. 4.3, the Lee—Schetzen method is used to determine the Wiener kernels (Fig. 4.3C). Here the kernels are used to predict the output by convolving the input with the kernels and adding up the contributions from each kernel (Fig. 4.3D). It can be seen that the predicted and recorded output match very well; this can be further confirmed when we compute the % variance that is accounted for (VAF) as:

VAF = (1-(std(zzw-est)^2)/(std(zzw)^2))*100

Here zzw and est are the measured and estimated output, respectively, and std is the MATLAB command to compute the standard deviation.

4.5   Relation between Wiener and Volterra Kernels

To summarize the preceding sections, a principal problem with the Volterra series is the dependence between the convolution-like terms (operators) in the series. This dependence prevents us from determining each term separately; this problem is resolved by Wiener's approach. To achieve the independence between terms, Wiener modified the individual terms in the series (Wiener operators are nonhomogeneous) and adapted the input (zero mean GWN). Volterra operators H_n have Volterra kernels (h_0, h_1, h_2, ...), whereas Wiener operators G_n have Wiener kernels (k_0, k_1, k_2, ...) as well as derived Wiener kernels ($k_{0(1)}$, $k_{0(2)}$, $k_{1(2)}$, ...).

Both Wiener and Volterra kernels are equivalent in the sense that the Wiener kernels can be determined from the Volterra kernels and vice versa. In our examples above we considered the zero- to the second-order kernels; let us assume that we are

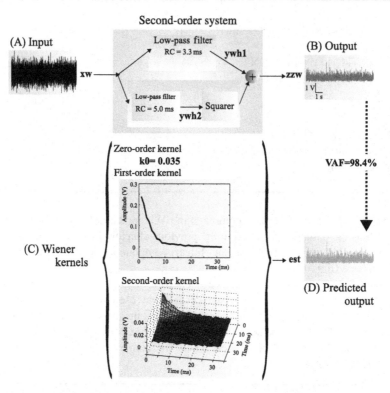

Figure 4.3 Wiener kernels of a second-order system similar to the one depicted in Fig. 3.2C; the example is computed with Pr4_2.m. (A) The input signal xw is GWN. (B) The output signal is zzw. (C) zero-, first- and second-order Wiener kernels computed by the MATLAB script using the procedure depicted in Fig. 4.2. (D) The predicted output est on the basis of the Wiener kernels approximates the measured output well: the variance accounted for VAF is 98.4%. (*Color in electronic version.*)

looking into a second-order system so that these are the only kernels available (all higher-order kernels are zero). In this case we have the following kernel components: k_0, $k_{0(1)}$, $k_{0(2)}$, k_1, $k_{1(2)}$, and k_2. In this example the Volterra kernels h_0-h_2 are:

$$h_0 = k_0 + k_{0(1)} + k_{0(2)} = k_0 + k_{0(2)}$$
$$h_1 = k_1 + k_{1(2)} = k_1 \qquad\qquad (4.27)$$
$$h_2 = k_2$$

The above equations for h_0-h_2 simplify because $k_{0(1)}$ and $k_{1(2)}$ are zero (see Equations (4.6) and (4.12)). So in a second-order system the relationship between the Wiener and Volterra kernels is fairly straightforward. Had we looked into a higher-order system, for example in a third-order system, we would add $k_{1(3)}$ to h_1 in Equation (4.27). The expressions for h_0 and h_2 remain unaltered because the other derived third-order kernels $k_{0(3)}$ and $k_{2(3)}$ are zero (Schetzen, 2006). Again, the rationale for this redistribution of kernel components is to create independence between the operators in the series (the condition in Equation (4.4d)). For example, by moving the term $k_{0(2)}$ from the zero-order expression (h_0) to the second-order Wiener operator, we satisfy the independence between the second-order Wiener operator and H_0 (Equation (4.8)). Considering the relationships in Equation (4.27), it is unsurprising that a comparison between our findings for the Wiener kernels k_1 and k_2, obtained with Pr4_2.m (depicted in Fig 4.3C), and the Volterra kernels h_1 and h_2, found in Pr3_2.m from the previous chapter, reveals a close resemblance.

From this chapter we can deduce that to obtain the Volterra kernels, we must know the system's order as well as all the Wiener kernels. In an experimental situation one usually does not know the system's order; at best one could estimate the order by establishing the number of Wiener kernels required to (sufficiently) approximate the system's output signal. In most experimental studies the Wiener kernels (up to the second or third order) and their contributions to the system's output are determined without any further attempt to identify the Volterra kernels.

4.6 Analyzing Spiking Neurons Stimulated with Noise

When studying intracellular or extracellular recordings of a spiking neuron while stimulating it with noise, one might (of course) use the raw output trace (including the action potentials) and relate this to the input as we have done previously (Fig. 4.2). However, instead of using the neuron's raw output signal, we can use alternative methods to represent the action potential activity. In the following discussion we assume that timing is the only relevant information associated with a neuronal spiking event. Methods that consider only spike timing can be applied to both intracellular and extracellular recordings of single cells. When dealing with high levels of spike activity it is common to represent the cell's output as the instantaneous spike rate (defined as (interspike interval)$^{-1}$) plotted vs. time; this procedure is shown in Fig. 1.1. Another frequently used technique is to bin the spike train and plot the number of spikes per bin against the time-stamp of the bin. However, if spike rates are low, both of these methods are impractical because we obtain time series that are either extremely unevenly sampled or too sparsely populated with values other than zeros and ones. In general, if one is interested only in the spike train, it seems reasonable to present the output time series of N spikes occurring at times t_i as a series of delta functions, thereby ignoring small subthreshold fluctuations of the neuronal response or noise in the recordings (chapter 14 in van Drongelen, 2007).

With a little bit of work, the Schetzen correlation method can be adapted to analyze spiking neurons stimulated by GWN. An example for the auditory system was described by Recio-Spinoso et al. (2005). In this study, the auditory system is stimulated by auditory noise and the authors represent the neuron's output y (a spike train of N spikes) as a series of Diracs at times t_i:

$$y(t) = \sum_{i=1}^{N} \delta(t - t_i) \tag{4.28}$$

Following our result in Equation (4.17), the **zero-order Wiener kernel** is the time average of the system's output:

$$k_0 = \langle y(t) \rangle = \left\langle \sum_{i=1}^{N} \delta(t - t_i) \right\rangle \tag{4.29a}$$

The time average $\langle \ldots \rangle$ can be written as an integral over the interval $[0,T]$, divided by epoch length T (that is $(1/T) \int_0^T \cdots$) :

$$\frac{1}{T} \int_0^T \sum_{i=1}^{N} \delta(t - t_i) \mathrm{d}t = \frac{1}{T} \sum_{i=1}^{N} \int_0^T \delta(t - t_i) \mathrm{d}t \tag{4.29b}$$

Here we interchanged the integration and summation operation. The timing t_i for each spike i is between 0 and T, so consequently the Dirac $\delta(t - t_i)$ is located within the $[0,T]$ integration interval and the integral $\int_0^T \delta(t - t_i) \mathrm{d}t$ evaluates to 1 (see Section 2.2.2 in van Drongelen, 2007). Therefore, the expression in Equation (4.29) simply counts the number N of action potentials divided by the time epoch T. Thus the zero-order Wiener kernel evaluates to N/T, which is the neuron's mean firing rate N_0:

$$k_0 = \frac{N}{T} = N_0 \tag{4.30}$$

The **first-order Wiener kernel** is given by Equation (4.20):

$$k_1(\tau_1) = \frac{1}{\sigma^2} \langle y(t) x(t - \tau_1) \rangle \tag{4.31}$$

If we rewrite the time average $\langle \ldots \rangle$ as an integral and substitute the output z in Equation (4.20) with the spike time series y (given in Equation (4.28)), we get:

$$k_1(\tau_1) = \frac{1}{\sigma^2} \left[\overbrace{\frac{1}{T} \int_0^T \underbrace{\left(\sum_{i=1}^{N} \delta(t - t_i) \right)}_{\text{output}} \underbrace{x(t - \tau_1)}_{\text{input}} \mathrm{d}t}^{\text{Time average}} \right]$$

$$= \frac{1}{\sigma^2} \left[\frac{1}{T} \int_0^T \left(\sum_{i=1}^{N} \delta(t - t_i) x(t - \tau_1) \right) \mathrm{d}t \right] \tag{4.32}$$

In the above we included input x in the summation. Now we again interchange the summation and integration operations:

$$k_1(\tau_1) = \frac{1}{\sigma^2}\left[\frac{1}{T}\sum_{i=1}^{N}\underbrace{\int_{0}^{T}\delta(t-t_i)x(t-\tau_1)\mathrm{d}t}_{x(t_i-\tau_1)}\right] = \frac{1}{\sigma^2}\underbrace{\frac{1}{T}N}_{N_0}\underbrace{\frac{1}{N}\sum_{i=1}^{N}x(t_i-\tau_1)}_{R_1(\tau_1)} \tag{4.33}$$

Here we evaluated the integral using the sifting property and multiplied the expression by N/N to allow substitution of $R_1(\tau_1)$, **the reverse-correlation function** (see section 14.5 in van Drongelen, 2007). The reverse-correlation function is also known as the revcor, which can be determined by averaging the stimulus time course that precedes each spike (spike-triggered average). If we think of the zero-order kernel as the time average (mean firing rate) of the system's output, we can conceptualize the first-order Wiener kernel as the average stimulus value some time τ_1 before spike i occurs (i.e., $x(t_i - \tau_1)$). Simplifying notation, we finally get:

$$\boxed{k_1(\tau_1) = \frac{N_0}{\sigma^2}R_1(\tau_1)} \tag{4.34}$$

The **second-order Wiener kernel**, on the other hand, represents the mean of the product of two copies of the input x (at two times τ_1 and τ_2) before the occurrence of a spike. The second-order Wiener kernel as given by Equation (4.24) becomes:

$$k_2(\tau_1,\tau_2) = \frac{1}{2\sigma^4}\langle y(t)x(t-\tau_1)x(t-\tau_2)\rangle \tag{4.35}$$

In Recio-Spinoso et al. (2005), the above equation is corrected by subtracting the zero-order kernel k_0 from the output. This makes sense for the following reasons. As discussed above, subtracting the contribution of lower-order kernels from the output is common practice (Fig. 4.2). In Equation (4.32) we did not correct the output for the first-order kernel estimate because theoretically its contribution should be independent from the zero-order one ($k_{0(1)}$ is zero, Equation (4.6)). However, we do correct for the DC (constant) term in the second-order estimate because a nonzero zero-order component $k_{0(2)}$ does exist (see Equation (4.9)). We will not correct y for the first-order contribution to k_2 because theoretically $k_{1(2)}$ is zero (Equation (4.12)). Therefore, y in Equation

(4.35) can simply be corrected for the zero-order contribution N_0 by using the output y minus the zero-order kernel:

$$y(t) - k_0 = \sum_{i=1}^{N} \delta(t - t_i) - N_0 \qquad (4.36)$$

By doing this we get:

$$k_2(\tau_1, \tau_2) = \frac{1}{2\sigma^4} \left\langle \left[\sum_{i=1}^{N} \delta(t - t_i) - N_0 \right] x(t - \tau_1)x(t - \tau_2) \right\rangle \qquad (4.37a)$$

Writing the time average in the integral notation, we get:

$$= \frac{1}{2\sigma^4} \left\{ \frac{1}{T} \int_0^T \left[\sum_{i=1}^{N} \delta(t - t_i) - N_0 \right] x(t - \tau_1)x(t - \tau_2)dt \right\} \qquad (4.37b)$$

We can write the expression as two separate integral terms:

$$= \frac{1}{2\sigma^4} \left\{ \frac{1}{T} \int_0^T \sum_{i=1}^{N} \delta(t - t_i)x(t - \tau_1)x(t - \tau_2)dt - \frac{1}{T} \int_0^T N_0 x(t - \tau_1)x(t - \tau_2)dt \right\} \qquad (4.37c)$$

By changing the integration and summation order in the **first term** and applying the sifting property for the Dirac, we get the following expression for the first term:

$$\frac{1}{2\sigma^4} \frac{1}{T} \sum_{i=1}^{N} \int_0^T \delta(t - t_i)x(t - \tau_1)x(t - \tau_2)dt = \frac{1}{2\sigma^4} \frac{1}{T} \sum_{i=1}^{N} \underbrace{x(t_i - \tau_1)x(t_i - \tau_2)}$$

$$(4.38a)$$

As we did with the first-order kernel earlier, we can multiply by N/N to simplify notation by using the expression for the **second-order reverse correlation** $R_2(\tau_1, \tau_2) = (1/N) \sum_{i=1}^{N} x(t_i - \tau_1)x(t_i - \tau_2)$. Finally, the first term in Equation (4.37c) simplifies to:

$$\frac{1}{2\sigma^4} \underbrace{\frac{1}{T} N}_{N_0} \underbrace{\frac{1}{N} \sum_{i=1}^{N} x(t_i - \tau_1)x(t_i - \tau_2)}_{R_2(\tau_1,\tau_2)} = \frac{N_0}{2\sigma^4} R_2(\tau_1, \tau_2) \qquad (4.38b)$$

The **second term** in Equation (4.37c) is:

$$= \frac{1}{2\sigma^4}\frac{1}{T}\int_0^T -N_0 x(t-\tau_1)x(t-\tau_2)dt$$

$$= -\frac{N_0}{2\sigma^4}\frac{1}{T}\underbrace{\int_0^T x(t-\tau_1)x(t-\tau_2)dt}_{\phi(\tau_2-\tau_1)} = -\frac{N_0}{2\sigma^4}\phi(\tau_2-\tau_1) \tag{4.39}$$

The expression $(1/T)\int_0^T x(t-\tau_1)x(t-\tau_2)dt$ is the autocorrelation $\phi(\tau_2-\tau_1)$ of the input noise. Note that unlike the variable t_i (representing the spike times) in the expression for the reverse correlation R_2, the time variable t is continuous in ϕ. Combining the results for the first and second terms we finally get:

$$k_2(\tau_1,\tau_2) = \frac{N_0}{2\sigma^4}[R_2(\tau_1,\tau_2) - \phi(\tau_2-\tau_1)] \tag{4.40a}$$

The above approach was used by Recio-Spinoso et al. (2005) to determine the first- and second-order Wiener kernels of different types of auditory nerve fibers. An example of the second-order kernel for a so-called low-characteristic frequency nerve fiber is shown in Fig. 4.4. If the input is zero mean GWN, we have $\phi(\tau_2-\tau_1) = \sigma^2\delta(\tau_2-\tau_1)$. Then, because we decided to ignore values where $\tau_1 = \tau_2$, as we did in Equation (4.24), we get:

$$k_2(\tau_1,\tau_2) = \frac{N_0}{2\sigma^4}R_2(\tau_1,\tau_2) \quad \text{for } \tau_1 \neq \tau_2 \tag{4.40b}$$

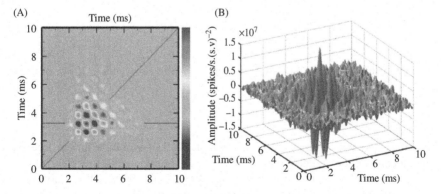

Figure 4.4 Example of a second-order kernel of an auditory low-characteristic frequency nerve fiber. (A) A 2D color-coded presentation of k_2. (B) The corresponding 3D plot of k_2. (*Panel A color in electronic version.*) (*From Recio-Spinoso et al., 2005*)

4.7 Nonwhite Gaussian Input

Zero mean GWN was selected as input signal for the determination of the Wiener series. In real applications, however, this is not feasible because the bandwidth of the noise is limited. In some cases, the bandwidth of the noise at the input may be wide enough relative to the bandwidth that is relevant for the system under investigation that we may consider the noise as white. However, there are situations where such an assumption is not valid. In these cases the input noise is band-limited (colored). The effect of using colored noise as input will be analyzed and discussed in the following paragraphs.

Recall that in Equation (4.40a) we left the noise autocorrelation term $\varphi(\tau_2 - \tau_1)$ in the expression. In Equation (4.40b), under the condition that the input is zero mean GWN, we ignored the correlation term because $\sigma^2 \delta(\tau_2 - \tau_1)$ evaluates to zero for $\tau_1 \neq \tau_2$. In general, when we consider systems, the noise presented at the input may be zero mean and Gaussian, but nonwhite (Gaussian colored noise [GCN]). The term "white" indicates that all frequencies are equally present in the noise signal, while in colored noise not all frequencies are equally present (i.e., we are dealing with filtered white noise). The filter effect has a direct consequence on the autocorrelation of the noise input (Fig. 4.5A and B). However, both colored and white noise may be Gaussian, a property that is related to their amplitude distribution (Fig. 4.5C and D).

In the following we assume that we have determined the zero-order kernel as the mean output and that we deal only with demeaned signals for input and output. Under this assumption, the cross-correlation φ_{xz} between GCN input x and output z can be developed similar to the procedure shown in Equations (4.18)−(4.20):

$$
\begin{aligned}
\phi_{xz}(v_1) = \langle z(t)x(t - v_1) \rangle &= \sum_{n=0}^{N} \langle G_n[k_n; x(t)]x(t - v_1) \rangle \\
&= \int_{-\infty}^{\infty} k_1(\tau_1)\langle x(t - \tau_1)x(t - v_1) \rangle \mathrm{d}\tau_1 = \int_{-\infty}^{\infty} k_1(\tau_1)\phi_{xx}(\tau_1 - v_1)\mathrm{d}\tau_1
\end{aligned}
\tag{4.41}
$$

The above shows that the cross-correlation ϕ_{xz} is the convolution of the first-order kernel k_1 with the input autocorrelation φ_{xx}:

$$
\phi_{xz} = k_1 \otimes \phi_{xx}
\tag{4.42a}
$$

Therefore k_1 can be obtained from the associated deconvolution. In the frequency domain convolution and deconvolution can be simplified to multiplication and division, respectively (section 8.3.2 in van Drongelen, 2007). The equivalent of Equation (4.42a) in the frequency domain therefore is:

$$
\Phi_{xz} = K_1 \Phi_{xx} \rightarrow K_1 = \Phi_{xz}/\Phi_{xx}
\tag{4.42b}
$$

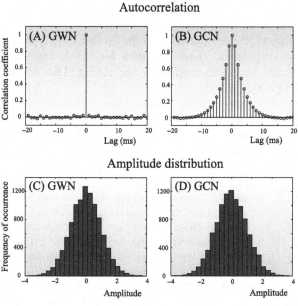

Figure 4.5 Autocorrelations (A and B) and amplitude distributions (C and D) of sampled Gaussian noise signals. The cases for GWN are depicted in (A) and (C). The same properties for colored (filtered) noise are shown in (B) and (D). (*Color in electronic version.*)

Here Φ_{XZ}, Φ_{XX}, and K_1 are the Fourier transforms of ϕ_{XZ}, ϕ_{XX}, and k_1, respectively. Now recall that the cross- and autocorrelation in the frequency domain can also be expressed as products (section 8.4.2 in van Drongelen, 2007) X^*Z and X^*X (where X and Z are the Fourier transforms of x and z, respectively, and * indicates the complex conjugate). Substituting these expressions for cross- and autocorrelation we get:

$$K_1 = X^*Z / X^*X \qquad (4.42c)$$

In real applications we can use this expression to determine K_1 by averaging Φ_{XZ} (X^*Z) and Φ_{XX} (X^*X) for each frequency f over a number of epochs:

$$K_1(f) = \langle X(f)^* Z(f) \rangle / \langle X(f)^* X(f) \rangle \qquad (4.42d)$$

Here the angle brackets $\langle \ldots \rangle$ indicate the average procedure in the frequency domain. Note the similarities and differences between this expression and the one for coherence (section 8.5 in van Drongelen, 2007). The inverse Fourier transform of K_1 in Equation (4.42d) gives k_1 for a nonlinear system with GCN input. A similar development for the second-order kernel gives us:

$$K_2(f_1, f_2) = \frac{\langle X(f_1)^* X(f_2)^* Z(f_1 + f_2) \rangle}{2 \langle X(f_1)^* X(f_1) \rangle \langle X(f_2)^* X(f_2) \rangle} \qquad (4.43)$$

and taking the inverse Fourier transform of the above expression then gives k_2.

4.8 Summary

As we demonstrated in this chapter, the determination of the Wiener kernels can be obtained from input—output correlation. These scenarios for GWN input are depicted in Fig. 4.6, both for the case with continuous output (Fig. 4.6A and B) and for spike train outputs (Fig. 4.6C—F).

Panels (A) and (B) in Fig. 4.6 show first- and second-order correlation procedures: the multiplication of $z(t)x(t - \tau_1)$ and $z(t)x(t - \tau_1)x(t - \tau_2)$, respectively. Because the cross-correlations are determined by the integration of these products, one may envision moving the multiplications over the signal while summing

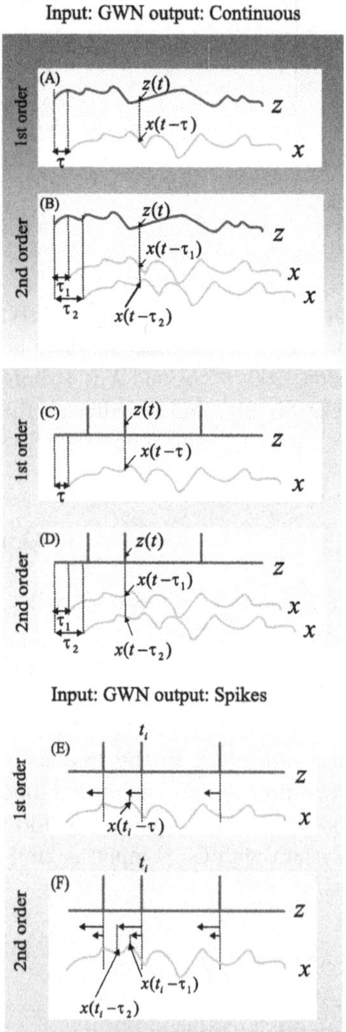

Figure 4.6 Summary diagrams of the characterization of a system with GWN input. Diagrams of the cross-correlation procedures for systems with a continuous output (A, B) or spike train output (C—F). (A), (C), and (E) show the first-order case and (B), (D), and (F) represent the second-order procedure. (C) and (E) depict two alternative visualizations for obtaining the first-order cross-correlation for systems with spiking output. In (C), the input is shifted by amount τ, whereas in (E), $x(t - \tau)$ at time $t = t_i$ is directly determined without shifting x (represented by the left-pointing arrow). For the spike output case, this procedure in (E) can be followed (as an alternative to the standard procedure in C) since the cross-correlation product is zero when there is no spike at time t_i. The analogous alternatives for determining the second-order correlation is shown in (D) and (F). See text for further explanation. (*Color in electronic version.*)

(integrating) the resulting products. The delays τ, τ_1, τ_2 can be visualized by shifting input x relative to output z (Fig. 4.6A and B).

If the system's output z is a spike train, as shown in (C) and (D), the correlations required to compute the kernels are identical: that is, the input can be shifted relative to the output to obtain $x(t - \tau)$, $x(t - \tau_1)$, and $x(t - \tau_2)$. However, this procedure can also be depicted as reverse correlations of each spike at time t_i, as shown in (E) and (F). Instead of shifting the input as we have just depicted, the reverse-correlation procedure is shown here with left-pointing arrows. Note that this is just another way of representing the shifts τ, τ_1, τ_2, and that it is not essentially different from the visualization in (C) and (D). However, the fact that we only consider the products $z(t)x(t - \tau)$ and $z(t)x(t - \tau_1)x(t - \tau_2)$ at t_i is essentially different from the case when we have a system with continuous output (as depicted in (A) and (B)) and is caused by the fact that we model the spike train with a series of Diracs. In between the spikes (i.e., in between the unit-impulse functions), the output $z(t)$ is considered zero and the products $z(t)x(t - \tau)$ and $z(t)x(t - \tau_1)x(t - \tau_2)$ vanish.

From the examples in this and the previous chapter, it may be clear that computing the kernels in the series can be a demanding task computationally. Recently, Franz and Schölkopf (2006) described an alternative method to estimate Volterra and Wiener series. Details of their approach are beyond the scope of this text, but the essence is to consider discrete systems only (which is not really a limitation if one wants to compute the series). In this case, the Volterra or Wiener series operators can be replaced by functions for which the parameters (the kernel parameters) can be estimated with regression techniques (see Section 2.4.1 for an example of a regression procedure). This approach is computationally more efficient than the Lee and Schetzen (1965) cross-correlation method (described here in Sections 4.3 and 4.4) and makes the estimation of high-order kernels feasible. An example of an application of this method to EEG is described in Barbero et al. (2009).

Appendix 4.1

Averages of Gaussian Random Variables

In this appendix we discuss averages of GWN variables because their properties are important for the development of the Wiener series approach (especially in demonstrating that the operators are orthogonal to lower-order operators). Because it is beyond the scope of this text to provide a detailed proof of all properties presented here, for further background see appendix A of Schetzen (2006). The relationship between higher- and lower-order moments, which we will discuss below, is also known as Wick's theorem (see, e.g., Zinn-Justin, 2002).

Let us consider ergodic and zero mean GWN represented by variable x: that is, the expected value of x can be replaced by its time average, which is zero (zero mean):

$$E\{x(t - \tau)\} = \langle x(t - \tau) \rangle = 0 \tag{A4.1.1}$$

The product $\langle x(t - \tau_1)x(t - \tau_2)\rangle$ is equal to the autocorrelation and also to the auto-covariance (because the noise is zero mean):

$$\langle x(t - \tau_1)x(t - \tau_2)\rangle = \sigma^2 \delta(\tau_1 - \tau_2) \tag{A4.1.2}$$

Because this may not be immediately apparent, let us define $t - \tau_1 = T$ and $\tau_2 - \tau_1 = \tau$. We can now rewrite the autocorrelation in Equation (A4.1.2) as $\langle x(T)x(T - \tau)\rangle$. In the case where $\tau = 0$ ($\tau_2 = \tau_1$), we get the expression $\langle x(T)x(T)\rangle = \langle x(T)^2\rangle = E\{x(T)^2\}$. For GWN with zero mean, this is the definition of the variance σ^2 of signal x (see section 3.2 in van Drongelen, 2007, on statistical moments). Again, since we are dealing with GWN (which gives us a random signal x), two different samples of x are uncorrelated; that is, for $\tau \neq 0$ ($\tau_2 \neq \tau_1$), $x(T)$ is not correlated with $x(T - \tau)$. This means that:

$$\langle x(T)x(T - \tau)\rangle = E\{x(T)x(T - \tau)\} = 0 \quad \text{for } \tau \neq 0.$$

Combining the above findings for $\tau_2 = \tau_1$ and $\tau_2 \neq \tau_1$ we can use the expression in Equation (A4.1.2) with the Dirac delta function. Let us look into an example in which we scale the correlation coefficient between ± 1. A scatter plot showing correlation for a GWN signal is shown in Fig. A4.1.1. The plot of the signal against itself with zero lag ($\tau = 0$) is depicted in Fig. A4.1.1A and obviously all points lie on the $y = x$ line, corresponding to a correlation coefficient of one. An example for a delay of $\tau = 1$ is shown in Fig. A4.1.1B; here the points are distributed in all directions corresponding to the absence of correlation (correlation coefficient of zero). This behavior is confirmed in a plot of the autocorrelation of GWN: we have a correlation coefficient of one for a lag τ of zero and a correlation coefficient of zero otherwise (see also Fig. 4.5A).

The findings from the paragraph above can be generalized to evaluate higher-order products between GWN signals (Schetzen, 2006). All averages of **odd** products evaluate to zero (e.g., $\langle x(t - \tau_1)x(t - \tau_2)x(t - \tau_3)\rangle = 0$), while it can be shown

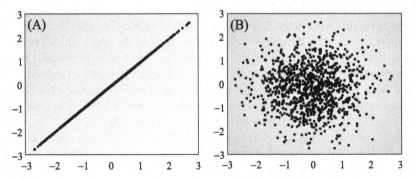

Figure A4.1.1 Correlation for $y(t)$, a digitized GWN signal of 1000 points. (A) A plot of $y(t)$ vs. $y(t)$. (B) A plot of $y(t + 1)$ vs. $y(t)$. (*Color in electronic version.*)

that higher-order **even** products are equal to the sum of all distinct pair-wise products. For example:

$$\langle x(t-\tau_1)x(t-\tau_2)x(t-\tau_3)x(t-\tau_4)\rangle = \langle x(t-\tau_1)x(t-\tau_2)\rangle\langle x(t-\tau_3)x(t-\tau_4)\rangle$$
$$+ \langle x(t-\tau_1)x(t-\tau_3)\rangle\langle x(t-\tau_2)x(t-\tau_4)\rangle + \langle x(t-\tau_1)x(t-\tau_4)\rangle\langle x(t-\tau_2)x(t-\tau_3)\rangle$$

$$(A4.1.3)$$

If you are interested in the formal proof of the above generalizations for the odd and even products, please see Appendix A in Schetzen (2006).

Appendix 4.2

Delay System as Volterra Operator

We used a specific delay operator earlier for creating the Hilbert transform in Chapter 1. Here we will comment on delay operators in general. Creation of a delay v_1 in $x(t)$ is an operation by which we obtain $x(t - v_1)$; this operation can be considered a 1D, first-order Volterra operator (Fig. A4.2.1A). Higher-dimensional (2D and 3D) delay systems can be represented by second- and third-order Volterra systems (Fig. A4.2.1B and C), etc. The 1D operator D_1 can be characterized by the notation:

$$D_1[x(t)] = x(t - v_1) \tag{A4.2.1}$$

Because this is a first-order system, this operation can be represented by a convolution:

$$D_1[x(t)] = x(t - v_1) = \int_{-\infty}^{\infty} h(\tau)x(t - \tau)\,\mathrm{d}\tau \tag{A4.2.2}$$

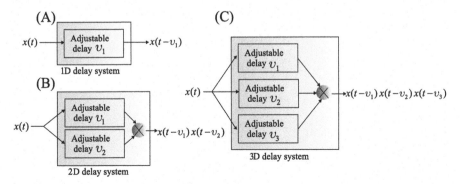

Figure A4.2.1 Examples of delay systems as Volterra operators. (*Color in electronic version.*)

From Equation (A4.2.2) we may conclude that the weighting function (the UIR) of the 1D system is $h(\tau) = \delta(\tau - v_1)$, thus resulting in:

$$x(t - v_1) = \int_{-\infty}^{\infty} \delta(\tau - v_1)x(t - \tau)d\tau \tag{A4.2.3}$$

Similarly, the delay operators for 2D operator D_2 and 3D operator D_3 can be defined as $D_2[x(t)] = x(t - v_1)x(t - v_2)$ and $D_3[x(t)] = x(t - v_1)x(t - v_2)x(t - v_3)$, respectively. In Fig. A4.2.1B and C we can see that each of the delays in the higher-dimensional system is a first-order operator. In the second-order (2D) system the UIRs are $\delta(\tau - v_1)$ and $\delta(\tau - v_2)$; in the third-order delay system, the UIRs are $\delta(\tau - v_1)$, $\delta(\tau - v_2)$, and $\delta(\tau - v_3)$. Similar to Equation (A4.2.3), these operations can be represented with the convolution-like integrals of the Volterra series (see Equation (3.4)); for example, in the 2D case:

$$x(t - v_1)x(t - v_1) = \int_{-\infty}^{\infty} \int_{-\infty}^{\infty} \underbrace{\delta(\tau_1 - v_1)\delta(\tau_2 - v_2)}_{\text{2nd order Volterra kernel } h_2(v_1, v_2)} x(t - \tau_1)x(t - \tau_2)d\tau_1\, d\tau_2$$

$$\tag{A4.2.4}$$

where the second-order Volterra kernel is:

$$h_2(v_1, v_2) = \delta(\tau_1 - v_1)\delta(\tau_2 - v_2) \tag{A4.2.5}$$

In the 3D case the third-order Volterra kernel for a delay system is:

$$h_3(v_1, v_2, v_3) = \delta(\tau_1 - v_1)\delta(\tau_2 - v_2)\delta(\tau_3 - v_3) \tag{A4.2.6}$$

Similarly, we can extend this approach to an n-dimensional delay operator:

$$h_n(v_1, v_2, \ldots, v_n) = \delta(\tau_1 - v_1)\delta(\tau_2 - v_2)\ldots\delta(\tau_n - v_n) \tag{A4.2.7}$$

5 Poisson−Wiener Series

5.1 Introduction

In the previous chapter we considered systems with continuous input signals. One such continuous input is Gaussian white noise (GWN), which allows us to create a series with orthogonal terms that can be estimated sequentially with the Lee−Schetzen cross-correlation method (also shown in the previous chapter). This approach can be adapted when the system's natural input consists of impulse trains such as a spike train. Identifying a system with an impulse train as input will be the topic of this chapter. We will elaborate on the approach that was described by Krausz (1975) and briefly summarized in Marmarelis (2004).[1] Our task at hand is to develop a Wiener series-like approach that describes the input−output relationship of a nonlinear system when an impulse train is at its input. To create randomness at the input, we use an impulse sequence that follows a Poisson process (see Section 14.2 in van Drongelen, 2007).

5.2 Systems with Impulse Train Input

The approach is to create a set of operators that are orthogonal to all lower-order Volterra operators, which is analogous to the development of the Wiener series with a GWN input. We will call these operators "Poisson−Wiener operators" to distinguish our current development of operators (using impulses as input) from that of Chapter 4 (using GWN as input). For each order n, we will symbolize these Poisson−Wiener operators as P_n. Similar to the Wiener series, we define the output z of a nonlinear system as the sum of a set of these operators, each depending on kernel p_n and impulse train input x. For a system of order N we have:

$$z(t) = \sum_{n=0}^{N} P_n[p_n; x(t)]$$

This equation for the Poisson−Wiener series is similar to the ones for the Volterra and Wiener series, but as we will see there are important differences.

As we described in the previous chapter, the approach of the Wiener series works so well because of the specific characteristics of the GWN input signal: $\langle x(t - \tau_1) \rangle = 0$, $\langle x(t - \tau_1)x(t - \tau_2) \rangle = \sigma^2 \delta(\tau_2 - \tau_1)$, etc. (see Appendix 4.1). When

---

[1] If you compare the following with Krausz' original work, please note that the derivation in Krausz (1975) contains minor scaling errors (as was also noted by Marmarelis, 2004).

Signal Processing for Neuroscientists, A Companion Volume. DOI: 10.1016/B978-0-12-384915-1.00005-X

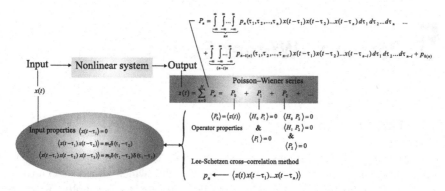

Figure 5.1 Diagram of the procedures used here to develop the Poisson—Wiener series, the properties of its operators, and the method to determine the kernels. Just as for the Wiener series, the input signal's properties play a crucial role in the development of the Poisson—Wiener approach. *(Color in electronic version.)*

the system's input changes to a series of impulses, these relationships no longer hold and we can no longer apply the equations we derived previously. To resolve this, we must start from scratch and first determine expressions for the averaged products $\langle x(t - \tau_1) \rangle$, $\langle x(t - \tau_1)x(t - \tau_2) \rangle$, ... for the Poisson process. Subsequently we must use these new results to redo the Gram—Schmidt procedure for the derivation of our series' orthogonal terms, as was done in Section 4.2. Finally we must redevelop Lee—Schetzen's cross-correlation method in a similar fashion as the procedure described in Section 4.3.

A schematic overview of the procedures we develop in this chapter is depicted in Fig. 5.1. Similar to the properties of Wiener series, the output z of a nonlinear system can be described by a (Poisson—Wiener) series in which:

(1) Operators P_n are heterogeneous (top-right in Fig. 5.1)
(2) Each operator is orthogonal to all lower-order Volterra operators
(3) Except for P_0, the Expectation (or time average) of all operators will vanish
(4) Except for p_0, the kernels can be determined from the cross-correlation of input and output (see also the Lee—Schetzen method introduced in Chapter 4).

In each of the above properties, it is important to know the Expectation or time average for the input and its cross products (see Input properties in Fig. 5.1). Therefore we will first determine these time averages associated with the input in Section 5.2.1 before we derive the Poisson—Wiener kernels in Section 5.2.2 and adapt Lee—Schetzen's cross-correlation method for determining the kernels from recorded data in Section 5.3.

5.2.1 Product Averages for the Poisson Impulse Train

Let us use signal χ, a train of Diracs with amplitude A that follows a Poisson process with rate ρ (Fig. 5.2A). The **first moment** or mean μ of impulse train χ can be established by a time average over a sufficiently long interval T. In such

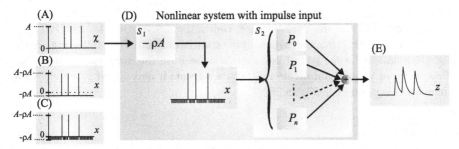

Figure 5.2 Impulse train inputs following a Poisson process can be used to identify nonlinear systems. A standard impulse train χ with amplitude A is shown in (A). A demeaned version of this time series x is depicted in (B). The signal in (C) is the same demeaned series x but is now presented as a series of weighted unit impulses (each impulse is represented by a vertical line). The procedure depicted in (D) shows the steps we use to identify a nonlinear system with such a train of impulses. First we pretend that the input is demeaned by part of the system (subsystem S_1) by subtracting ρA, the mean of χ. This demeaned series x is then used as input to subsystem S_2. We actually determine the operators P_n and kernels for S_2 instead of the whole system $S_1 + S_2$, but if we can characterize S_2 we have characterized the whole system, since S_1 is a simple subtraction. (E) depicts the output z of the system to the impulse input. *(Color in electronic version.)*

an interval we expect to find $N = \rho T$ impulses in the input signal $\chi = \sum\limits_{i=1}^{N=\rho T}$ $A\delta(t - t_i)$. The time average of the input signal is $\langle \chi \rangle = (1/T) \int\limits_0^T \sum\limits_{i=1}^{N=\rho T} A\delta(t - t_i)\mathrm{d}t$. Assuming we can interchange summation and integration we get:

$$\mu = \langle \chi \rangle = \frac{A}{T} \sum_{i=1}^{N=\rho T} \underbrace{\int_0^T \delta(t - t_i)\mathrm{d}t}_{=1} = \frac{A}{T}\rho T = \rho A \tag{5.1a}$$

The integral in Equation (5.1a) evaluates to one if the delta function is located within the interval T (i.e., $0 \le t_i \le T$). We could have done the computation of the mean in a simpler way because we know how many impulses we expect in epoch T and the amplitude of each impulse. The number of impulses (each with amplitude A) during this interval is ρT, resulting in the following expression for the first moment:

$$\mu = \langle \chi \rangle = \frac{1}{T} \int_0^T \underbrace{\rho T}_{\text{I}} \underbrace{A\delta(t)}_{\text{II}} \ \mathrm{d}t = \frac{1}{T}\rho TA = \rho A \tag{5.1b}$$

Part I in the equation above is the number of expected impulses over interval T and Part II is the amplitude for each impulse. Unlike the first moment for the GWN signal we used in the previous chapter, this result is **not** zero. The following step is therefore critical for the rest of our approach: **because the nonzero result for the first moment would complicate matters, we create a new signal x, which is the demeaned version of χ** (Fig. 5.2B):

$$\boxed{x(t) = \chi(t) - \rho A} \tag{5.1c}$$

We can check that this generates a zero first moment for time series $x(t)$:

$$\langle x \rangle = \frac{1}{T} \int_0^T \left[\sum_{i=1}^{N=\rho T} A\delta(t - t_i) - \rho A \right] dt = \frac{1}{T} \left(\sum_{i=1}^{N=\rho T} \underbrace{\int_0^T A\delta(t - t_i)dt}_{A} - \int_0^T \rho A\, dt \right) \tag{5.1d}$$

$$= \frac{1}{T}\left(\rho AT - [\rho At]_0^T \right) = \frac{1}{T}(\rho AT - \rho AT) = 0$$

Here we interchanged the integration and summation operations. Subsequently, we evaluate the integral with the delta function and find that it is equal to the constant A if the delta function falls within epoch T. Alternatively, we can also approach the estimation of $\langle x \rangle$ a bit differently. As you can see in Fig. 5.2C, we can consider the demeaned signal as a series of Diracs (a sampled version of the signal) with amplitude $A - A\rho$ for each spike, and amplitude $-A\rho$ in between the spikes. Over interval T the number of spike occurrences is again ρT and the number of nonspike occurrences is $(1 - \rho)T$.

$$\langle x \rangle = \frac{1}{T} \int_0^T [\underbrace{\rho T}_{\text{I}}\, \underbrace{(A - A\rho)\delta(t)}_{\text{II}} + \underbrace{(1 - \rho)T}_{\text{III}}\, \underbrace{(-A\rho)\delta(t)}_{\text{IV}}]dt \tag{5.1e}$$

$$= \frac{1}{T} \int_0^T \underbrace{[\rho AT - \rho^2 AT - \rho AT + \rho^2 AT]}_{0} \delta(t)dt = 0$$

Parts I and III above are the expected number of spiking and nonspiking events, respectively, and Parts II and IV are their respective amplitudes. We will use the approach in Equation (5.1e) to compute the higher-order products in the following. The bottom line is that by using the impulse series x as input, we have (just as for GWN) zero for the first moment m_1:

$$\boxed{m_1 = \langle x \rangle = 0} \tag{5.1f}$$

The next expression we must evaluate is the cross-correlation $\langle x(t - \tau_1)x(t - \tau_2)\rangle$. To start, we can look into the *second moment* $\langle x^2 \rangle$ of the impulse train in Fig. 5.2C. As shown above in Equation (5.1e), the number of events N is the event probability ρ times the interval T, and the nonevent probability equals $(1 - \rho)T$ (Parts I and III, respectively). For the second-order moment, we will square the associated amplitudes (Parts II and IV):

$$\langle x^2 \rangle = \frac{1}{T}\int_0^T [\underbrace{\rho T}_{\text{I}}\ \underbrace{(A - A\rho)^2 \delta(t)}_{\text{II}}\ +\ \underbrace{(1 - \rho)T}_{\text{III}}\ \underbrace{(-A\rho)^2 \delta(t)}_{\text{IV}}]dt \qquad (5.2a)$$

Note that by squaring the amplitudes we weight the unit impulse function $\delta(t)$, but we do not need to square the delta function itself. It is relatively simple to see why this is not required. Imagine the input as the series of Dirac deltas weighted with different amplitudes shown in Fig. 5.2C. The sum of all amplitudes x divided by the epoch length T is the first moment, the sum of all x^2 divided by T is the second moment, the sum of all x^3 divided by T is the third moment, and so on (see Section 3.2 in van Drongelen, 2007). To sample the amplitudes of x, x^2, x^3, ... we only have to weight a single Dirac with the desired amplitude (if you need to review the properties of the Dirac δ, see section 2.2.2 in van Drongelen, 2007). Simplifying Equation (5.2a), we get:

$$= \frac{1}{T}\int_0^T [\underbrace{\rho A^2 T - 2\rho^2 A^2 T + \rho^3 A^2 T + \rho^2 A^2 T - \rho^3 A^2 T]}_{\rho A^2 T - \rho^2 A^2 T}\ \delta(t)dt$$

$$= \frac{1}{T}\int_0^T T\rho A^2 (1 - \rho)\delta(t)dt = \frac{1}{T}\left[T\rho A^2(1 - \rho)\right]$$

Finally, the expression for the second moment m_2 becomes:

$$m_2 = \langle x^2 \rangle = \rho A^2(1 - \rho) \qquad (5.2b)$$

The next step is to determine the second-order cross-correlation using a time average of the product $x(t - \tau_1)x(t - \tau_2)$:

$$\langle x(t - \tau_1)x(t - \tau_2)\rangle = \frac{1}{T}\int_0^T [\underbrace{\rho T}_{\text{I}}\ \underbrace{(A - A\rho)^2 \delta(t - \tau_1)\delta(t - \tau_2)}_{\text{II}}$$
$$+ \underbrace{(1 - \rho)T}_{\text{III}}\ \underbrace{(-A\rho)^2 \delta(t - \tau_1)\delta(t - \tau_2)}_{\text{IV}}]dt \qquad (5.3a)$$

Parts I–IV are similar to the ones in Equation (5.2a). The product of I and II, the first term in the integral in Equation (5.3a), evaluates to:

$$\frac{1}{T}\left[\rho T(A - A\rho)^2\right]\delta(\tau_1 - \tau_2)$$

and the product of III and IV, the second term in Equation (5.3a), becomes:

$$\frac{1}{T}\left[(1 - \rho)TA^2\rho^2\right]\delta(\tau_1 - \tau_2)$$

Combining the two terms above, we get the result for the *second-order cross-correlation*:

$$\boxed{\langle x(t - \tau_1)x(t - \tau_2)\rangle = \rho A^2(1 - \rho)\delta(\tau_1 - \tau_2) = m_2\delta(\tau_1 - \tau_2)} \qquad (5.3b)$$

This result is not unexpected since Equation (5.3b) becomes the expression we derived for the second moment m_2 (Equation (5.2b)) when we have the case $\tau_1 = \tau_2$. Just as was the case for GWN, this expression will evaluate to zero otherwise.

For computing the *third moment* m_3, we can use the same approach as in Equation (5.2a):

$$\langle x^3 \rangle = \frac{1}{T}\int_0^T [\underbrace{\rho T}_{\text{I}}\ \underbrace{(A - A\rho)^3\delta(t)}_{\text{II}}\ +\ \underbrace{(1 - \rho)T}_{\text{III}}\ \underbrace{(-A\rho)^3\delta(t)}_{\text{IV}}]dt \qquad (5.4a)$$

If you do the algebra, you will find that this results in:

$$m_3 = \langle x^3 \rangle = \rho A^3(1 - \rho)(1 - 2\rho) \qquad (5.4b)$$

The third-order cross-correlation is:

$$\langle x(t - \tau_1)x(t - \tau_2)x(t - \tau_3)\rangle$$

$$= \frac{1}{T}\int_0^T [\underbrace{\rho T}_{\text{I}}\ \underbrace{(A - A\rho)^3\delta(t - \tau_1)\delta(t - \tau_2)\delta(t - \tau_3)}_{\text{II}}$$

$$+ \underbrace{(1 - \rho)T}_{\text{III}}\ \underbrace{(-A\rho)^3\delta(t - \tau_1)\delta(t - \tau_2)\delta(t - \tau_3)}_{\text{IV}}]dt \qquad (5.5a)$$

in which Parts I−IV can be evaluated similarly to the ones in Equation (5.3a). Accordingly, the result becomes:

$$\langle x(t - \tau_1)x(t - \tau_2)x(t - \tau_3) \rangle = \rho A^3 (1 - \rho)(1 - 2\rho)\delta(\tau_1 - \tau_2)\delta(\tau_1 - \tau_3)$$

Combined with Equation (5.4b), we get:

$$\boxed{\langle x(t - \tau_1)x(t - \tau_2)x(t - \tau_3) \rangle = m_3 \delta(\tau_1 - \tau_2)\delta(\tau_1 - \tau_3)} \tag{5.5b}$$

for the *third-order cross-correlation*. As you can see, due to the presence of two Diracs, the third-order product is only nonzero for $\tau_1 = \tau_2 = \tau_3$.

In the above cases, things are relatively simple because we set the first moment to zero by demeaning the input impulse train. This approach ensures that any product that contains $E\{x\}$ or $\langle x \rangle$ (the Expectation or time average of x) vanishes (see Appendix 5.1). Appendix 5.1 explains that for the *fourth moment* m_4, we have to deal with additional terms that include $E\{x^2\}$. If you are mainly interested in how we will next make Poisson−Wiener operators orthogonal, you can accept the results for m_4 and the fourth-order product below and skip the appendix. The expression for m_4 is obtained in the same manner as the lower-order moments above.

$$m_4 = \langle x^4 \rangle = \rho A^4 [\rho(1 - \rho)^2 + (1 - \rho)(1 - 2\rho)^2] \tag{5.6a}$$

The time averaged *fourth-order cross-correlation* critically depends on the values of the delays $\tau_1 - \tau_4$ in a piece-wise manner (see Appendix 5.1, Equation (A5.1.5)):

$$
\begin{aligned}
&\langle x(t - \tau_1)x(t - \tau_2)x(t - \tau_3)x(t - \tau_4) \rangle \\[4pt]
&\begin{cases}
\tau_1 = \tau_2 = \tau_3 = \tau_4 & : m_4 \delta(\tau_1 - \tau_2)\delta(\tau_1 - \tau_3)\delta(\tau_1 - \tau_4) \\
\tau_1 = \tau_2 \text{ and } \tau_3 = \tau_4 & : m_2^2 \delta(\tau_1 - \tau_2)\delta(\tau_3 - \tau_4) \\
\tau_1 = \tau_3 \text{ and } \tau_2 = \tau_4 & : m_2^2 \delta(\tau_1 - \tau_3)\delta(\tau_2 - \tau_4) \\
\tau_1 = \tau_4 \text{ and } \tau_2 = \tau_3 & : m_2^2 \delta(\tau_1 - \tau_4)\delta(\tau_2 - \tau_3) \\
0 & \text{otherwise}
\end{cases}
\end{aligned}
\tag{5.6b}
$$

5.2.2 Orthogonal Terms of the Poisson−Wiener Series

In this section we use the same procedure (Gram−Schmidt orthogonalization, see Arfken and Weber, 2005) as in Chapter 4 to derive the orthogonal series that can characterize a nonlinear system given our impulse input. As depicted in Fig. 5.2D, the Poisson−Wiener series represents an output signal z consisting of the sum of operators P_n:

$$z(t) = P_0[p_0; x(t)] + P_1[p_1; x(t)] + P_2[p_2; x(t)] + \cdots + P_n[p_n; x(t)] \tag{5.7a}$$

in which the heterogeneous operator P_n is defined as:

$$
\begin{aligned}
& P_n[p_n; x(t)] \\
&= \underbrace{\int_{-\infty}^{\infty} \int_{-\infty}^{\infty} \cdots \int_{-\infty}^{\infty}}_{n\times} p_n(\tau_1, \tau_2, \ldots, \tau_n) x(t-\tau_1) x(t-\tau_2) \ldots x(t-\tau_n) d\tau_1\, d\tau_2 \ldots d\tau_n \\
&\quad + \underbrace{\int_{-\infty}^{\infty} \int_{-\infty}^{\infty} \cdots \int_{-\infty}^{\infty}}_{(n-1)\times} p_{n-1(n)}(\tau_1, \tau_2, \ldots, \tau_{n-1}) x(t-\tau_1) x(t-\tau_2) \ldots x(t-\tau_{n-1}) d\tau_1\, d\tau_2 \ldots d\tau_{n-1} \\
&\quad + \underbrace{\int_{-\infty}^{\infty} \int_{-\infty}^{\infty} \cdots \int_{-\infty}^{\infty}}_{(n-i)\times} p_{n-i(n)}(\tau_1, \tau_2, \ldots, \tau_{n-i}) x(t-\tau_1) x(t-\tau_2) \ldots x(t-\tau_{n-i}) d\tau_1\, d\tau_2 \ldots d\tau_{n-i} + p_{0(n)}
\end{aligned}
$$

$$(5.7b)$$

Here we have Poisson−Wiener kernel p_n and derived Poisson−Wiener kernels $p_{n-i(n)}$ ($i = 1, 2, \ldots, n$). In Sections 5.2.2.1−5.2.2.3, we will derive the expressions for the Poisson−Wiener operators in a similar fashion we did for the Wiener series in Chapter 4.

5.2.2.1 The Zero-Order Poisson−Wiener Operator

Similar to the zero-order Wiener operator, we define the zero-order Poisson−Wiener operator P_0 as the output's DC component p_0:

$$\boxed{P_0[p_0; x(t)] = p_0}$$

$$(5.8)$$

In this equation, we use p_0 to symbolize the zero-order Poisson−Wiener kernel to distinguish it from the zero-order Volterra and Wiener kernels h_0 and k_0, respectively.

5.2.2.2 The First-Order Poisson−Wiener Operator

Now we use the orthogonality between Poisson−Wiener operators and lower-order Volterra operators to derive the expression for the first-order Poisson−Wiener kernel p_1. Similar to Equation (4.5), we have:

$$\langle H_0[x(t)]P_1[p_1;x(t)]\rangle = \left\langle h_0 \underbrace{\left[\int_{-\infty}^{\infty} p_1(\tau_1)x(t-\tau_1)d\tau_1 + p_{0(1)} \right]}_{P_1} \right\rangle = 0$$

$$= h_0 \underbrace{\left[\int_{-\infty}^{\infty} p_1(\tau_1)\langle x(t-\tau_1)\rangle d\tau_1 + p_{0(1)} \right]}_{\langle P_1 \rangle} = 0$$

(5.9)

The subscript 0(1) indicates that $p_{0(1)}$ is a derived kernel: a zero-order member of the first-order operator P_1. Note that we took all constants out of the time average operation, and only the (time dependent) input time series x remains within the time average brackets $\langle...\rangle$. Since input x is a demeaned impulse train following a Poisson process, we know that $\langle x(t-\tau_1)\rangle = 0$ (see Equation (5.1f)). Consequently the integral evaluates to zero, and we therefore conclude that the orthogonality requirement demands that:

$$\boxed{p_{0(1)} = 0}$$

(5.10)

Substituting this result in the general expression for our first-order Poisson–Wiener operator $P_1[p_1;x(t)] = \int_{-\infty}^{\infty} p_1(\tau_1)x(t-\tau_1)d\tau_1 + p_{0(1)}$, we obtain:

$$\boxed{P_1[p_1;x(t)] = \int_{-\infty}^{\infty} p_1(\tau_1)x(t-\tau_1)d\tau_1}$$

(5.11)

Note that this result is very similar to the first-order Wiener operator (Equation (4.7)). Furthermore, we see that $E\{P\} = \langle P_1 \rangle = 0$: that is, the Expectation or time average of P_1, $\langle \int_{-\infty}^{\infty} p_1(\tau_1)x(t-\tau_1)d\tau_1 \rangle$, evaluates to zero because $\langle x(t-\tau_1)\rangle = 0$.

You can also see in Fig. 5.2D that this kernel is not the first-order kernel for our system but for the subsystem indicated by S_2 (the whole system is $S_1 + S_2$). Because we know that the other part, subsystem S_1, is a simple subtraction ($-\rho A$), we have effectively characterized the first-order component of the system under investigation.

5.2.2.3 The Second-Order Poisson–Wiener Operator

To establish the expression for the second-order operator we follow the same procedure as for the Wiener kernels: we demand both orthogonality between the second-order Poisson–Wiener operator and a zero-order Volterra operator plus orthogonality between the second-order operator and a first-order Volterra operator.

First, for orthogonality between H_0 and P_2: using the orthogonality condition we get:

$$\langle H_0[x(t)]P_2[p_2;x(t)]\rangle = 0$$

That is:

$$= \left\langle \underbrace{h_0}_{H_0} \underbrace{\left[\int\limits_{-\infty}^{\infty}\int\limits_{-\infty}^{\infty} p_2(\tau_1,\tau_2)x(t-\tau_1)x(t-\tau_2)\mathrm{d}\tau_1\mathrm{d}\tau_2 + \int\limits_{-\infty}^{\infty} p_{1(2)}(\tau_1)x(t-\tau_1)\mathrm{d}\tau_1 + p_{0(2)} \right]}_{P_2} \right\rangle = 0$$

$$= h_0 \left[\int\limits_{-\infty}^{\infty}\int\limits_{-\infty}^{\infty} p_2(\tau_1,\tau_2)\langle x(t-\tau_1)x(t-\tau_2)\rangle\mathrm{d}\tau_1\mathrm{d}\tau_2 \right.$$

$$\left. + \int\limits_{-\infty}^{\infty} p_{1(2)}(\tau_1)\langle x(t-\tau_1)\rangle\mathrm{d}\tau_1 + p_{0(2)} \right] = 0 \tag{5.12}$$

Similar to the composition of the Wiener operator G_2, the components $p_{0(2)}$ and $p_{1(2)}$ are derived zero-order and first-order members of operator P_2. As we did in Equation (5.9), we took all constants out of the time average $\langle...\rangle$ and only kept the time series x within it. Again, because the input is a zero mean impulse train following a Poisson process, the term with the single integral in the expression above is zero (since $\langle x(t-\tau_1)\rangle = 0$, Equation (5.1f)). The term with the double integral is dictated by the averaged product of both inputs $\langle x(t-\tau_1)x(t-\tau_2)\rangle$, which is given by Equation (5.3b). Therefore the above expression becomes:

$$h_0 \int\limits_{-\infty}^{\infty}\int\limits_{-\infty}^{\infty} p_2(\tau_1,\tau_2)\langle x(t-\tau_1)x(t-\tau_2)\rangle\mathrm{d}\tau_1\,\mathrm{d}\tau_2 + h_0 p_{0(2)}$$

$$= m_2 h_0 \int\limits_{-\infty}^{\infty}\int\limits_{-\infty}^{\infty} p_2(\tau_1,\tau_2)\delta(\tau_1-\tau_2)\mathrm{d}\tau_1\,\mathrm{d}\tau_2 + h_0 p_{0(2)} = 0$$

This equation can be evaluated by using the sifting property for one of the time constants; here we integrate with respect to τ_2 and get:

$$m_2 h_0 \int\limits_{-\infty}^{\infty} p_2(\tau_1,\tau_1)\mathrm{d}\tau_1 + h_0 p_{0(2)} = 0 \rightarrow \boxed{p_{0(2)} = -m_2 \int\limits_{-\infty}^{\infty} p_2(\tau_1,\tau_1)\mathrm{d}\tau_1} \tag{5.13}$$

As you can see, the kernel $p_{0(2)}$ is derived from p_2.

To further express our second-order Poisson–Wiener operator, we will next demand orthogonality between second-order operator P_2 and first-order Volterra operator H_1. Similar to Equation (4.10), we have:

$$\langle H_1[x(t)]P_2[p_2; x(t)] \rangle = 0,$$

which can be written as:

$$\left\langle \left[\int_{-\infty}^{\infty} h_1(v)x(t-v)\mathrm{d}v \right] \times \left[\int_{-\infty}^{\infty} \int_{-\infty}^{\infty} p_2(\tau_1, \tau_2)x(t-\tau_1)x(t-\tau_2)\mathrm{d}\tau_1\,\mathrm{d}\tau_2 + \int_{-\infty}^{\infty} p_{1(2)}(\tau_1)x(t-\tau_1)\mathrm{d}\tau_1 + p_{0(2)} \right] \right\rangle = 0$$

(5.14)

Equation (5.14) contains three terms that we will consider separately. The **first** term is:

$$\left\langle \left[\int_{-\infty}^{\infty} h_1(v)x(t-v)\mathrm{d}v \right] \left[\int_{-\infty}^{\infty} \int_{-\infty}^{\infty} p_2(\tau_1, \tau_2)x(t-\tau_1)x(t-\tau_2)\mathrm{d}\tau_1\,\mathrm{d}\tau_2 \right] \right\rangle$$

$$= \int_{-\infty}^{\infty} \int_{-\infty}^{\infty} \int_{-\infty}^{\infty} h_1(v)p_2(\tau_1, \tau_2) \underbrace{\langle x(t-v)x(t-\tau_1)x(t-\tau_2) \rangle}_{m_3\delta(v-\tau_1)\delta(v-\tau_2)}\,\mathrm{d}v\,\mathrm{d}\tau_1\,\mathrm{d}\tau_2$$

In the Wiener series development, for systems with GWN input, the odd product $\langle x(t-v)x(t-\tau_1)x(t-\tau_2) \rangle = 0$. Here, however, the odd product is given by Equation (5.5b). This gives:

$$m_3 \int_{-\infty}^{\infty} \int_{-\infty}^{\infty} \int_{-\infty}^{\infty} h_1(v)p_2(\tau_1, \tau_2)\delta(v-\tau_1)\delta(v-\tau_2)\mathrm{d}v\,\mathrm{d}\tau_1\,\mathrm{d}\tau_2$$

(5.15a)

$$= m_3 \int_{-\infty}^{\infty} h_1(v)p_2(v, v)\mathrm{d}v$$

The **second** term in Equation (5.14) is:

$$\left\langle \left[\int_{-\infty}^{\infty} h_1(v)x(t-v)\mathrm{d}v \right] \left[\int_{-\infty}^{\infty} p_{1(2)}(\tau_1)x(t-\tau_1)\mathrm{d}\tau_1 \right] \right\rangle$$

$$= \int_{-\infty}^{\infty} \int_{-\infty}^{\infty} h_1(v)p_{1(2)}(\tau_1) \underbrace{\langle x(t-v)x(t-\tau_1) \rangle}_{m_2\delta(v-\tau_1)}\,\mathrm{d}v\,\mathrm{d}\tau_1$$

Using the expression for the second-order correlation in Equation (5.3b) we can simplify to:

$$m_2 \int_{-\infty}^{\infty} \int_{-\infty}^{\infty} h_1(v)p_{1(2)}(\tau_1)\delta(v-\tau_1)dv\,d\tau_1 = m_2 \int_{-\infty}^{\infty} h_1(v)p_{1(2)}(v)dv \qquad (5.15b)$$

Note that we used the sifting property of the Dirac to simplify the double integral.
    Finally, the **third** term in Equation (5.14) is:

$$\left\langle \left[\int_{-\infty}^{\infty} h_1(v)x(t-v)dv \right] p_{0(2)} \right\rangle = \left[\int_{-\infty}^{\infty} h_1(v)\langle x(t-v)\rangle dv \right] p_{0(2)} = 0 \qquad (5.15c)$$

This evaluates to zero because $\langle x(t-v)\rangle = 0$ (Equation (5.1f)).

    Substituting the results from Equations (5.15a), (5.15b), and (5.15c) into Equation (5.14), we have:

$$m_3 \int_{-\infty}^{\infty} h_1(v)p_2(v,v)dv + m_2 \int_{-\infty}^{\infty} h_1(v)p_{1(2)}(v)dv = 0$$

From this we may conclude that the derived first-order member of the second-order operator is:

$$\boxed{p_{1(2)}(v) = -\frac{m_3}{m_2}p_2(v,v)} \qquad (5.16)$$

Again, you can see that the derived kernel $p_{1(2)}$ is indeed derived because it fully depends on p_2. Using the results in Equations (5.13) and (5.16), we get the expressions for the second-order Poisson−Wiener operator in terms of the second-order Poisson−Wiener kernel p_2:

$$\begin{aligned}
P_2[p_2; x(t)] = &\int_{-\infty}^{\infty} \int_{-\infty}^{\infty} p_2(\tau_1,\tau_2)x(t-\tau_1)x(t-\tau_2)d\tau_1\,d\tau_2 \\
&\underbrace{-\frac{m_3}{m_2} \int_{-\infty}^{\infty} p_2(\tau,\tau)x(t-\tau)d\tau}_{+\int p_{1(2)}x(t-\tau)d\tau} \underbrace{- m_2 \int_{-\infty}^{\infty} p_2(\tau,\tau)d\tau}_{+p_{0(2)}}
\end{aligned} \qquad (5.17)$$

Note that the above result for the second-order Poisson–Wiener operator differs from the second-order Wiener operator (Equation (4.13)) (here $p_{1(2)}$ is nonzero). This difference is due to the fact that the cross-correlation results for a demeaned train of impulses following a Poisson process are different from a GWN signal (see Section 5.2.1 and compare Appendices 4.1 and 5.1). Using the expressions for $\langle x(t - \tau_1) \rangle$ and $\langle x(t - \tau_1)x(t - \tau_2) \rangle$, it is straightforward to show that $E\{P_2\} = \langle P_2 \rangle = 0$.

5.3 Determination of the Zero-, First- and Second-Order Poisson–Wiener Kernels

In this section we will compute the Poisson–Wiener kernels using the same cross-correlation method first described for the Wiener kernels (Lee and Schetzen, 1965). If we deal with a nonlinear system of order N, and we present a demeaned impulse train x following a Poisson process at its input, we obtain output z as the sum of the Poisson–Wiener operators (Fig. 5.2D):

$$z(t) = \sum_{n=0}^{N} P_n[p_n; x(t)] \tag{5.18}$$

In the following example we will describe how to determine the zero-, first- and second-order Poisson–Wiener kernels.

5.3.1 Determination of the Zero-Order Poisson–Wiener Kernel

Similar to the Wiener operators, the Expectation of all Poisson–Wiener operators P_n, except the zero-order operator P_0, is zero. Therefore, assuming an ergodic process (time averages are allowed for estimating the Expectations), we find the average of output signal z:

$$\boxed{\langle z(t) \rangle = \sum_{n=0}^{N} \langle P_n[p_n; x(t)] \rangle = P_0[p_0; x(t)] = p_0} \tag{5.19}$$

Thus the zero-order Poisson–Wiener kernel is equal to the mean output $\langle z(t) \rangle$.

5.3.2 Determination of the First-Order Poisson–Wiener Kernel

Similar to the procedure for the Wiener kernels depicted in Fig. 4.2, the first-order Poisson–Wiener kernel of a system can be obtained from the cross-correlation between its input and output:

$$\langle z(t)x(t-v_1)\rangle = \langle P_0[p_0;x(t)]x(t-v_1)\rangle + \langle P_1[p_1;x(t)]x(t-v_1)\rangle$$
$$+ \langle P_2[p_2;x(t)]x(t-v_1)\rangle + \cdots \tag{5.20}$$
$$= \sum_{n=0}^{N} \langle P_n[p_n;x(t)]x(t-v_1)\rangle$$

Recall that Poisson–Wiener kernels are defined to be orthogonal to lower-order Volterra kernels, and recall that the delay operator $x(t-v_1)$ can be presented as a first-order Volterra operator (see Appendix 4.2). Therefore, all Poisson–Wiener operators P_n with $n \geq 2$ are orthogonal to $x(t-v_1)$, and we only have to deal with operators of order $n = 0$ and 1.

For **n = 0**:

$$\langle P_0[p_0;x(t)]x(t-v_1)\rangle = p_0 \underbrace{\langle x(t-v_1)\rangle}_{0} = 0 \tag{5.21a}$$

For **n = 1**:

$$\langle P_1[p_1;x(t)]x(t-v_1)\rangle = \int_{-\infty}^{\infty} p_1(\tau_1) \underbrace{\langle x(t-\tau_1)x(t-v_1)\rangle}_{m_2\delta(\tau_1-v_1)} \, d\tau_1$$
$$= m_2 \int_{-\infty}^{\infty} p_1(\tau_1)\delta(\tau_1-v_1)d\tau_1 = m_2 p_1(v_1) \tag{5.21b}$$

Here we used Equation (5.3b) to simplify $\langle x(t-\tau_1)x(t-v_1)\rangle$ and then used the sifting property of the Dirac to evaluate the above integral. From the results in Equation (5.21b), we conclude that the only nonzero part in Equation (5.20) is the term for $n = 1$; therefore, the first-order Poisson–Wiener kernel becomes:

$$\langle z(t)x(t-v_1)\rangle = m_2 p_1(v_1) \rightarrow \boxed{p_1(v_1) = \frac{1}{m_2}\langle z(t)x(t-v_1)\rangle} \tag{5.22a}$$

Therefore, the first-order Poisson–Wiener kernel is the cross-correlation between input and output weighted by the second moment m_2 of the input.

We can use the properties of the Dirac to rewrite the cross-correlation expression, because the input is an impulse train. If we substitute the expression for the input in Equation (5.22a) with a sum of Diracs and present the time average $\langle ... \rangle$ with an integral notation $(1/T)\int_0^T \cdots$, we get:

$$p_1(v_1) = \frac{1}{m_2}\frac{1}{T}\int_0^T z(t)\overbrace{\left[A\sum_{i=1}^{N=\rho T}\delta(t-t_i-v_1) - \rho A\right]}^{\text{input: } x(t-v_1)} dt$$
$$\underbrace{\phantom{p_1(v_1) = \frac{1}{m_2}\frac{1}{T}\int_0^T z(t)\left[A\sum_{i=1}^{N=\rho T}\delta(t-t_i-v_1) - \rho A\right] dt}}_{\text{Time average}}$$

Assuming we may interchange the integration and summation and separating the terms for the impulse train (the Diracs) and the DC correction (ρA), this evaluates into two integral terms:

$$p_1(v_1) = \frac{A}{m_2} \frac{1}{T} \underbrace{\sum_{i=1}^{N=\rho T} \int_0^T z(t)\delta(t - t_i - v_1)dt}_{z(t_i + v_1)} - \frac{\rho A}{m_2} \frac{1}{T} \underbrace{\int_0^T z(t)dt}_{\langle z \rangle}$$

When using the sifting property it can be seen that the first term is a scaled average of $z(t_i + v_1)$ and may be rewritten as:

$$\frac{A}{m_2} \frac{\rho T}{T} \frac{1}{\rho T} \underbrace{\sum_{i=1}^{N=\rho T} z(t_i + v_1)}_{C_{zx}(v_1)} = \frac{\rho A}{m_2} C_{zx}(v_1) = \frac{\mu}{m_2} C_{zx}(v_1)$$

Note that we used the first moment $\mu = \rho A$ of the original train of impulses χ here (Equation (5.1a)). Combining the above we get:

$$\boxed{p_1(v_1) = \frac{\mu}{m_2}[C_{zx}(v_1) - \langle z \rangle]}$$

(5.22b)

The average $(1/\rho T) \sum_{i=1}^{N=\rho T} z(t_i + v_1)$ is the cross-correlation $C_{zx}(v_1)$ between the input impulse train x and the system's output z. **Unlike the reverse correlation we discussed in section 14.5 in van Drongelen (2007) and applied in Section 4.6, we deal with the forward-correlation here (see Fig. 5.5E).** In the examples in Chapter 4, we used reversed correlation because the impulse train was the output caused by the input and we had to go back in time to reflect this causality. In this case the role is reversed: the impulse train is the input causing the output.

5.3.3 Determination of the Second-Order Poisson–Wiener Kernel

Using a procedure analogous to that developed for the Wiener kernel in Section 4.3.3, we find the second-order Poisson–Wiener kernel by using a second-order cross-correlation between output and input:

$$\begin{aligned}
\langle z(t)x(t - v_1)x(t - v_2)\rangle &= \langle P_0[p_0; x(t)]x(t - v_1)x(t - v_2)\rangle \\
&+ \langle P_1[p_1; x(t)]x(t - v_1)x(t - v_2)\rangle \\
&+ \langle P_2[p_2; x(t)]x(t - v_1)x(t - v_2)\rangle + \cdots \\
&= \sum_{n=0}^{N} \langle P_n[p_n; x(t)]x(t - v_1)x(t - v_2)\rangle
\end{aligned}$$

(5.23)

Since $x(t - v_1)x(t - v_2)$ can be presented as a second-order Volterra operator (see Appendix 4.2), all Poisson–Wiener operators P_n with $n \geq 3$ are orthogonal to $x(t - v_1)x(t - v_2)$ (because all Poisson–Wiener operators are orthogonal to lower-order Volterra operators). Furthermore, since we use a Poisson process as input, we will not allow impulses to coincide. Therefore, we neglect all results for equal delays $v_1 = v_2$ in the evaluation of Equation (5.23). Taking into account the considerations above, we now analyze the second-order cross-correlation for $n = 0, 1, 2$, and $v_1 \neq v_2$.

For **$n = 0$**:

$$\langle P_0[p_0; x(t)]x(t - v_1)x(t - v_2)\rangle = p_0 \underbrace{\langle x(t - v_1)x(t - v_2)\rangle}_{m_2\delta(v_1 - v_2)} = m_2 p_0 \delta(v_1 - v_2)$$

(5.24a)

We can neglect this term because, due to the Dirac, it evaluates to zero for $v_1 \neq v_2$.

For **$n = 1$**:

$$\langle P_1[p_1; x(t)]x(t - v_1)x(t - v_2)\rangle = \int_{-\infty}^{\infty} p_1(\tau_1) \underbrace{\langle x(t - \tau_1)x(t - v_1)x(t - v_2)\rangle}_{m_3\delta(\tau_1 - v_1)\delta(\tau_1 - v_2)} \, d\tau_1$$

$$= m_3 p_1(v_1)\delta(v_1 - v_2) \qquad (5.24b)$$

Due to the Dirac, this expression also evaluates to zero for $v_1 \neq v_2$ and can therefore be ignored.

For **$n = 2$**, we compute $\langle P_2[p_2; x(t)]x(t - v_1)x(t - v_2)\rangle$ using Equation (5.17) and we get:

$$\overbrace{\int_{-\infty}^{\infty}\int_{-\infty}^{\infty} p_2(\tau_1, \tau_2) \underbrace{\langle x(t - \tau_1)x(t - \tau_2)x(t - v_1)x(t - v_2)\rangle}_{\text{Equation (5.6b)}} \, d\tau_1 \, d\tau_2}^{\text{I}}$$

$$\overbrace{-\frac{m_3}{m_2} \int_{-\infty}^{\infty} p_2(\tau_1, \tau_1) \underbrace{\langle x(t - \tau_1)x(t - v_1)x(t - v_2)\rangle}_{m_3\delta(\tau_1 - v_1)\delta(\tau_1 - v_2)} \, d\tau_1}^{\text{II}}$$

$$\overbrace{- m_2 \int_{-\infty}^{\infty} p_2(\tau_1, \tau_1) \underbrace{\langle x(t - v_1)x(t - v_2)\rangle}_{m_2\delta(v_1 - v_2)} \, d\tau_1}^{\text{III}} \qquad (5.24c)$$

Term I in Equation (5.24c) is the most complex one and potentially consists of four terms (Equation (5.6b)). Given that we have four delays τ_1, τ_2, v_1, v_2 and taking into account the condition $v_1 \neq v_2$, there are only two combinations that remain to be considered: $\tau_1 = v_1$ and $\tau_2 = v_2$ and $\tau_1 = v_2$ and $\tau_2 = v_1$. The first term can now be rewritten as:

$$\int\limits_{-\infty}^{\infty} \int\limits_{-\infty}^{\infty} p_2(\tau_1, \tau_2) \underbrace{\langle x(t - \tau_1)x(t - \tau_2)x(t - v_1)x(t - v_2) \rangle}_{m_2^2 \delta(\tau_1 - v_1)\delta(\tau_2 - v_2) + m_2^2 \delta(\tau_1 - v_2)\delta(\tau_2 - v_1)} d\tau_1 \, d\tau_2$$

$$= \int\limits_{-\infty}^{\infty} \int\limits_{-\infty}^{\infty} p_2(\tau_1, \tau_2)[m_2^2 \delta(\tau_1 - v_1)\delta(\tau_2 - v_2) + m_2^2 \delta(\tau_1 - v_2)\delta(\tau_2 - v_1)]d\tau_1 \, d\tau_2$$

$$= m_2^2 \underbrace{\int\limits_{-\infty}^{\infty} \int\limits_{-\infty}^{\infty} p_2(\tau_1, \tau_2)\delta(\tau_1 - v_1)\delta(\tau_2 - v_2)d\tau_1 \, d\tau_2}_{p_2(v_1, v_2)}$$

$$+ m_2^2 \underbrace{\int\limits_{-\infty}^{\infty} \int\limits_{-\infty}^{\infty} p_2(\tau_1, \tau_2)\delta(\tau_1 - v_2)\delta(\tau_2 - v_1)d\tau_1 \, d\tau_2}_{p_2(v_2, v_1)}$$

The double integral above can be evaluated by sifting for τ_1 and τ_2. Because we assume that the kernel is symmetric around its diagonal, we can use $p_2(v_1, v_2) = p_2(v_2, v_1)$, and the above evaluates to:

$$\boxed{2m_2^2 p_2(v_1, v_2)} \tag{5.24d}$$

Term II in Equation (5.24c) can be written as:

$$-\frac{m_3}{m_2} \int\limits_{-\infty}^{\infty} p_2(\tau_1, \tau_1)m_3 \delta(\tau_1 - v_1)\delta(\tau_1 - v_2)d\tau_1 = -\frac{m_3^2}{m_2}p_2(v_1, v_1)\delta(v_1 - v_2)$$

$$\tag{5.24e}$$

Due to the delta function $\delta(v_1 - v_2)$, this part can be neglected since it is zero for $v_1 \neq v_2$.

Term III in Equation (5.24c) evaluates to:

$$-m_2 \int\limits_{-\infty}^{\infty} p_2(\tau_1, \tau_1)m_2 \delta(v_1 - v_2)d\tau_1 = -m_2^2 \int\limits_{-\infty}^{\infty} p_2(\tau_1, \tau_1)\delta(v_1 - v_2)d\tau_1$$

$$\tag{5.24f}$$

This term can also be ignored because it equals zero for $v_1 \neq v_2$.

To summarize Equation (5.24), the only nonzero term for $v_1 \neq v_2$ is the result in Equation (5.24d). Substituting this result into Equation (5.23), we get an expression for our second-order Poisson−Wiener kernel p_2:

$$
\langle z(t)x(t - v_1)x(t - v_2)\rangle = 2m_2^2 p_2(v_1, v_2) \rightarrow
$$

$$
p_2(v_1, v_2) = \frac{1}{2m_2^2} \langle z(t)x(t - v_1)x(t - v_2)\rangle \quad \text{for } v_1 \neq v_2
$$

(5.25a)

Using the fact that the input x is a train of impulses, we can employ the same treatment as for Equation (5.22a) and rewrite Equation (5.25a) as:

$$
p_2(v_1, v_2) = \frac{1}{2m_2^2} \frac{1}{T} \int_0^T z(t) \overbrace{\left[A \sum_{i=1}^{N=\rho T} \delta(t - t_i - v_1) - \rho A \right]}^{\text{1st copy of the input: } x(t - v_1)} \times
$$

$$
\overbrace{\left[A \sum_{j=1}^{N=\rho T} \delta(t - t_j - v_2) - \rho A \right]}^{\text{2nd copy of the input: } x(t - v_2)} dt
$$

This expression generates four terms:

$$
\text{I:} \quad \frac{A^2}{2m_2^2} \frac{1}{T} \int_0^T z(t) \left[\sum_{i=1}^{N=\rho T} \delta(t - t_i - v_1) \right] \left[\sum_{j=1}^{N=\rho T} \delta(t - t_j - v_2) \right] dt
$$

$$
\text{II:} \quad -\frac{\rho A^2}{2m_2^2} \frac{1}{T} \int_0^T z(t) \sum_{i=1}^{N=\rho T} \delta(t - t_i - v_1) dt = \frac{\mu^2}{2m_2^2} C_{zx}(v_1), \text{ with:}
$$

$$
C_{zx}(v_1) = \frac{1}{\rho T} \sum_{i=1}^{N=\rho T} z(t_i + v_1) \text{ and } \mu = \rho A
$$

$$
\text{III:} \quad -\frac{\rho A^2}{2m_2^2} \frac{1}{T} \int_0^T z(t) \sum_{j=1}^{N=\rho T} \delta(t - t_j - v_2) dt = \frac{\mu^2}{2m_2^2} C_{zx}(v_2), \text{ with:}
$$

$$
C_{zx}(v_2) = \frac{1}{\rho T} \sum_{j=1}^{N=\rho T} z(t_j + v_2) \text{ and } \mu = \rho A
$$

$$
\text{IV:} \quad \frac{\rho^2 A^2}{2m_2^2} \frac{1}{T} \int_0^T z(t) dt = \frac{\mu^2}{2m_2^2} \langle z \rangle
$$

Terms II—IV were evaluated in a similar fashion as in the first-order case in Equation (5.22). After changing the order of the integration and summations, term I above evaluates to:

$$\frac{A^2}{2m_2^2}\frac{1}{T}\sum_{i=1}^{N=\rho T}\sum_{j=1}^{N=\rho T}\underbrace{\int_0^T z(t)[\delta(t-t_i-v_1)][\delta(t-t_j-v_2)]\mathrm{d}t}_{z(t_i+v_1)\delta(t_i-t_j+v_1-v_2)}$$

$$=\frac{A^2}{2m_2^2}\frac{\rho^2 T^2}{T}\frac{1}{\rho^2 T^2}\underbrace{\sum_{i=1}^{N=\rho T}\sum_{j=1}^{N=\rho T}z(t_i+v_1)\delta(t_i-t_j+v_1-v_2)}_{C_{zxx}(v_1,v_2)}=\frac{\mu^2}{2m_2^2}TC_{zxx}(v_1,v_2)$$

In the above expression we substituted μ for ρA (Equation (5.1a)); this is the first moment of the original impulse train χ. Combining the results from the four terms I—IV above, we have:

$$\boxed{p_2(v_1,v_2)=\frac{\mu^2}{2m_2^2}\{TC_{zxx}(v_1,v_2)-[C_{zx}(v_1)+C_{zx}(v_2)-\langle z\rangle]\}\ \text{ for }v_1\neq v_2}\qquad (5.25\mathrm{b})$$

The second-order correlation $C_{zxx}(v_1,v_2)$ is the average of $(1/\rho^2 T^2)$ $\sum_{i=1}^{N=\rho T}\sum_{j=1}^{N=\rho T}z(t_i+v_1)$ under the condition set by the Dirac $\delta(t_i-t_j+v_1-v_2)$. This condition is equivalent to sampling the values of output signal z when $t_i-t_j+v_1-v_2=0$. This indicates that:

(1) The delay between the copies of the input is $\Delta=v_2-v_1=t_i-t_j$, which means that the delays under consideration for creating the averages are equal to the differences Δ between spike times t_i,t_j.
(2) There is a relationship between the individual delays given by $v_2=t_i-t_j+v_1$, which represents a line in the v_1,v_2 plane at $45°$ and with an intercept at t_i-t_j.

This conditional average is therefore a slice through $p_2(v_1,v_2)$ defined by this line. The delays we consider are strictly given by t_i-t_j and the input to the averaging procedure is $z(t_i+v_1)$. A representation of C_{zxx} is shown in Fig. 5.5F. To keep Fig. 5.5F compatible with the symbols in the other panels in this figure, the delay v_1 is replaced by τ_1 in the diagram.

5.4 Implementation of the Cross-Correlation Method

Because there is no standard command in MATLAB to create a series of randomly occurring impulses following a Poisson process, we include an example function Poisson.m to create such an impulse train (for details see Appendix 5.2). In MATLAB script Pr5_1.m, we use this function to create the input (in

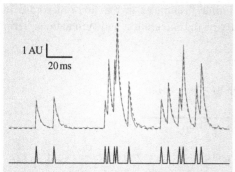

Figure 5.3 Example of input (pulses, lower line (black)) and output (dashed line (red)) traces. The (green) line, following the output closely, is the output contribution from the Poisson–Wiener kernels. The vertical scale is in arbitrary units (AU). The VAF by the model output in this example was 97.6%. All traces were generated by Pr5_1.m. *(Color in electronic version.)*

this example, impulses with amplitude of 2 units) to a nonlinear system consisting of a first-order component (a low-pass filter) and a second-order component (a low-pass filter amplifier with 5× amplification plus a squarer), similar to the system in Fig. 3.2C. Typical traces for input and output are shown in Fig. 5.3. By following the same steps depicted in Fig. 4.2 for the Wiener kernels, but now using Equations (5.19), (5.22b), and (5.25b), we find the Poisson–Wiener kernels for the system. Note that the cross-correlations are impulse-triggered averages in this case.

*The following MATLAB code is part of script **Pr5_1.m** and shows the computation of the first-order cross-correlation and first-order kernel p1 according to Equation (5.22b) (Step 3 of the Lee–Schetzen method depicted in Fig. 4.2).*

```
% Step 3. Create the first order average (see Fig. 4.2)
%
Czx=zeros(T,1);
for  i=1:length(time)-10              % to avoid problems by ignoring last
                                      % 10 impulses
      Czx=Czx+v0(time(i):time(i)+T-1);
end;
% Now we scale Czx by the # of spikes (i.e. length(time) − 10, which is the
% # of trials in the average. Using Equation (5.22b):
p1=(u1/m2)*((Czx/(length(time)-10))-mean(v0));   % Note that all scaling
                                                 % parameters
                                                 % u1, m2, and mean(v0)
                                                 % are at the ms - scale !
figure;
plot(p1)
title('first order Poisson-Wiener kernel')
xlabel('Time (1 ms)')
ylabel('Amplitude')
```

The percentage of variance accounted for (VAF, see Section 4.4 for its definition) by the output from the Poisson–Wiener kernels in this example is typically in the high 90 s. This VAF number is fairly optimistic because, as can be seen in the output trace in Fig. 5.3, a large number of points with a good match between output (dashed red line) and predicted output (green line) are zero or close to zero; the predicted output we mainly care about is (of course) the activity caused by the input (impulses) and not the rest state.

5.5 Spiking Output

In Chapter 4, we considered continuous input to nonlinear systems with both continuous and spiking output. So far in this chapter, we have analyzed nonlinear systems with spike train input and continuous output. The possible cases one might encounter in neuroscience are summarized in Fig. 5.4. As you can see, the only case remaining for our discussion is a nonlinear system with both spike input and output (Fig. 5.4D). We can compute the Poisson–Wiener kernels by using the previously found expressions (Equations (5.19), (5.22b), and (5.25b)). In this case, kernel p_0 can be determined by the time average of the spike output. Just as in Equation (4.30) p_0 evaluates to the mean firing rate of the output. In Equations (5.22b) and (5.25b)

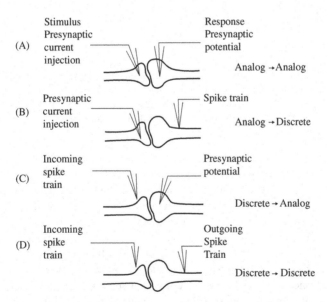

Figure 5.4 Types of signals one may encounter for a system's input and output in neuroscience. In this example a synapse is used to symbolize the four different possibilities (A–D). The incoming signal may be a GWN signal (analog–presynaptic current injection) or a train of impulses following a Poisson process (discrete–incoming spike train). The output can be a postsynaptic potential (analog) or a spike train (discrete). *(Fig. 11.1 from Marmarelis and Marmarelis (1978). With kind permission of Springer Science and Business Media.)*

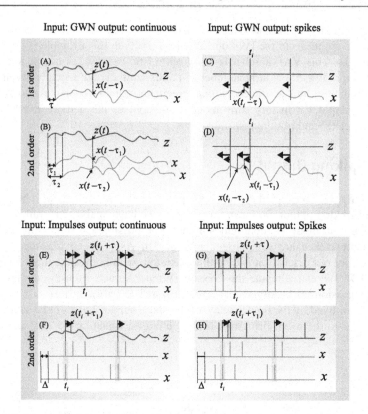

Figure 5.5 Summary of the procedures for determining first- and second-order cross-correlation for the different scenarios depicted in Fig. 5.4. In all panels, x (green) is the input and z (red) is the output. The panels for GWN input (A–D) are identical to Fig. 4.6A, B, E, F. Panels E–H show the procedures for impulses as input. See text for further explanation. *(Color in electronic version.)*

we can see that computing first- and second-order kernels require first- and second-order cross-correlations C_{zx} and C_{zxx} (in this case spike-triggered averages). The procedures for obtaining these cross-correlations are depicted in Fig. 5.5G and H. The first-order cross-correlation is a spike-triggered average; we use the input spikes as the trigger (Fig. 5.5G). The second-order cross-correlation is triggered by coinciding spikes of two copies from the input, one of which is shifted by amount Δ (Fig. 5.5H). The procedures for obtaining these cross-correlation functions are very similar to the ones discussed for a system with a spike input and continuous output, as you can see by comparing panels E with G and F with H in Fig. 5.5.

5.6 Summary

The procedures for determining the first- and second-order cross-correlations for the four scenarios in Fig. 5.4 are summarized in Fig. 5.5. The part of this figure for GWN input is identical to the overview in Fig. 4.6. The panels for spike input show the procedures discussed in this chapter. In practice, the cross-correlations required for computation of the Poisson–Wiener kernels can all be obtained from spike-triggered averages (Fig. 5.5E–H). As such it is very similar to the procedure we followed for nonlinear systems with GWN input and spike output in Chapter 4 (Fig. 5.5C and D). The difference is that here we use the input spikes, instead of the output spike train, to trigger the average; hence, we determine forward cross-correlation instead of reversed correlation. This reflects that the systems are considered causal (output is caused by input). Thus, a system's output shows reversed correlation with the input (Fig. 5.5C and D) and its input is forward-correlated with its output (Fig. 5.5E–H). The procedures followed to obtain the cross-correlations for systems with both continuous input and output are depicted in Fig. 5.5A and B. Here the correlation products are not spike-triggered and the delays of the copies of the input are determined for each sample of the output $z(t)$ (Chapter 4).

Appendix 5.1

Expectation and Time Averages of Variables Following a Poisson Process

The results for time averages of GWN are well known and were briefly summarized in Appendix 4.1. For the application of impulse trains we use a different input signal, the Poisson process (see section 14.2 in van Drongelen, 2007). Products of variables following a Poisson process are important for determining the Poisson–Wiener kernels when impulse trains are used as input to a nonlinear system. A similar derivation was described by Krausz (1975) in his appendix A.[2] Assuming that $x(t)$ follows a Poisson process, we can define the **first moment** as the Expectation of x: $E\{x\}$. Because the signal is ergodic, we may replace this with a time average $\langle x \rangle = (1/T) \int_0^T x(t) \mathrm{d}t$ (see section 3.2 in van Drongelen, 2007, if you need to review ergodicity and time averages). To simplify things further down the road, we start from a demeaned impulse train so that (see Equation (5.1)):

$$\boxed{E\{x\} = \langle x \rangle = 0} \tag{A5.1.1}$$

The Expectation of the **second-order product**, or cross-correlation, of variable x is $E\{x(t - \tau_1)x(t - \tau_2)\}$ (for cross-correlation, see section 8.4 in van Drongelen, 2007). Note that the expression we use here is slightly different from Equation (8.13) in van Drongelen (2007): we substituted $t - \tau_1$ and $t - \tau_2$ for t_1 and t_2, respectively. Because x follows a Poisson process, the factors $x(t - \tau_1)$ and

---

[2] Please note that the derivation by Krausz contains minor errors for the moments m_2 and m_3, leading to differences in the scaling of several of the derived expressions.

$x(t - \tau_2)$ are independent if $\tau_1 \neq \tau_2$; in this case we may replace the Expectation with two separate ones—that is:

$$E\{x(t - \tau_1)x(t - \tau_2)\} = E\{x(t - \tau_1)\}E\{x(t - \tau_2)\} = 0 \quad \text{for } \tau_1 \neq \tau_2$$

The above product evaluates to zero, because the first moment of our impulse train is zero. The expression $E\{x(t - \tau_1)x(t - \tau_2)\}$ is only nonzero if $\tau_1 = \tau_2$, and (again) because x is ergodic we may apply a time average $\langle x(t - \tau_1)x(t - \tau_2)\rangle$. In Section 5.2.1 you can see that the final result for the Expectation/time average of the second-order product becomes:

$$\boxed{E\{x(t - \tau_1)x(t - \tau_2)\} = \langle x(t - \tau_1)x(t - \tau_2)\rangle = m_2\delta(\tau_1 - \tau_2)} \qquad \text{(A5.1.2)}$$

The Expectation of the **third-order product** $E\{x(t - \tau_1)x(t - \tau_2)x(t - \tau_3)\}$ equals zero by independence if $\tau_1 \neq \tau_2 \neq \tau_3$, since in this case we can rewrite the expression as:

$$E\{x(t - \tau_1)x(t - \tau_2)x(t - \tau_3)\} = E\{x(t - \tau_1)\}E\{x(t - \tau_2)\}E\{x(t - \tau_3)\} = 0$$
$$\text{for } \tau_1 \neq \tau_2 \neq \tau_3$$

If only one pair of τ's is equal (i.e., $\tau_1 = \tau_2 \neq \tau_3$ or $\tau_1 \neq \tau_2 = \tau_3$), we can make the substitutions $\tau_1 = \tau_2$ or $\tau_2 = \tau_3$ and then separate the Expectation into two factors:

$$\begin{aligned} E\{x(t - \tau_1)x(t - \tau_2)x(t - \tau_3)\} &= E\{x(t - \tau_1)x(t - \tau_1)x(t - \tau_3)\} \\ &= E\{x(t - \tau_1)^2 x(t - \tau_3)\} \qquad \text{for } \tau_1 = \tau_2 \neq \tau_3 \\ &= E\{x(t - \tau_1)^2\}E\{x(t - \tau_3)\} = 0 \end{aligned}$$

and,

$$E\{x(t - \tau_1)x(t - \tau_2)x(t - \tau_3)\} = E\{x(t - \tau_2)^2\}E\{x(t - \tau_1)\} = 0 \quad \text{for } \tau_1 \neq \tau_2 = \tau_3$$

In all of the above cases, the expressions evaluate to zero because $E\{x\} = 0$, and the only instance where the Expectation of the third-order product is nonzero is for $\tau_1 = \tau_2 = \tau_3$. In this case (due to ergodicity), it may be replaced by $\langle x(t - \tau_1)x(t - \tau_2)x(t - \tau_3)\rangle$ (see Equation (5.5b)). The final nonzero result is:

$$\boxed{\begin{aligned} E\{x(t - \tau_1)x(t - \tau_2)x(t - \tau_3)\} &= \langle x(t - \tau_1)x(t - \tau_2)x(t - \tau_3)\rangle \\ &= m_3\delta(\tau_1 - \tau_2)\delta(\tau_1 - \tau_3) \end{aligned}} \qquad \text{(A5.1.3)}$$

The Expectation of the **fourth-order product** $E\{x(t - \tau_1)x(t - \tau_2)x(t - \tau_3)x(t - \tau_4)\}$ is zero by independence if:

I. $\tau_1 \neq \tau_2 \neq \tau_3 \neq \tau_4$

and nonzero if all delays are equal:

II. $\tau_1 = \tau_2 = \tau_3 = \tau_4$

Using the time average approach we use in Section 5.2, we find the following for the fourth moment:

$$m_4 = \langle x^4 \rangle = \frac{1}{T} \int_0^T [\rho T(A - A\rho)^4 \delta(t) + (1 - \rho)T(-A\rho)^4 \delta(t)]dt$$

This can be written as:

$$m_4 = \langle x^4 \rangle = \rho A^4 (1 - 4\rho + 6\rho^2 - 3\rho^3) = \rho A^4 [\rho(1 - \rho)^2 + (1 - \rho)(1 - 2\rho)^2]$$

Including the condition $\tau_1 = \tau_2 = \tau_3 = \tau_4$, we find that the averaged product is:

$$\boxed{\langle x(t - \tau_1)x(t - \tau_2)x(t - \tau_3)x(t - \tau_4) \rangle = m_4 \delta(\tau_1 - \tau_2)\delta(\tau_1 - \tau_3)\delta(\tau_1 - \tau_4)}$$

(A5.1.4)

in which the δ functions represent the condition that all delays must be equal for a nonzero result. Three alternatives with three equal delays are:

III. $\tau_1 \neq \tau_2 = \tau_3 = \tau_4$
IV. $\tau_1 = \tau_2 \neq \tau_3 = \tau_4$
V. $\tau_1 = \tau_2 = \tau_3 \neq \tau_4$

In all three cases **III–V**, the Expectation of the fourth-order product evaluates to zero. For instance in case **V** we have:

$$E\{x(t - \tau_1)x(t - \tau_2)x(t - \tau_3)x(t - \tau_4)\} = E\{x(t - \tau_1)^3 x(t - \tau_4)\}$$

$$= \underbrace{E\{x(t - \tau_1)^3\}}_{m_3} \underbrace{E\{x(t - \tau_4)\}}_{0} = 0$$

for $\tau_1 = \tau_2 = \tau_3 \neq \tau_4$

Finally, we have three cases in which delays are equal in pairs:

VI. $\tau_1 = \tau_2$ and $\tau_3 = \tau_4$
VII. $\tau_1 = \tau_3$ and $\tau_2 = \tau_4$
VIII. $\tau_1 = \tau_4$ and $\tau_2 = \tau_3$

These cases evaluate to a nonzero value. For instance, in case **VI** we get:

$$E\{x(t - \tau_1)x(t - \tau_2)x(t - \tau_3)x(t - \tau_4)\} = E\{x(t - \tau_1)^2 x(t - \tau_3)^2\}$$

$$= \underbrace{E\{x(t - \tau_1)^2\}}_{m_2} \underbrace{E\{x(t - \tau_3)^2\}}_{m_2} = m_2^2$$

$$\text{for } \tau_1 = \tau_2 \text{ and } \tau_3 = \tau_4$$

If we represent the conditions $\tau_1 = \tau_2$ and $\tau_3 = \tau_4$, with Dirac delta functions, we get the final result for case **VI**:

$$m_2^2 \delta(\tau_1 - \tau_2)\delta(\tau_3 - \tau_4)$$

To summarize the results for the Expectation of the fourth-order product:

$$
\begin{array}{l}
E\{x(t - \tau_1)x(t - \tau_2)x(t - \tau_3)x(t - \tau_4)\} \\
\left\{
\begin{array}{ll}
\tau_1 = \tau_2 \text{ and } \tau_3 = \tau_4 & : m_4 \delta(\tau_1 - \tau_2)\delta(\tau_1 - \tau_3)\delta(\tau_1 - \tau_4) \\
\tau_1 = \tau_2 \text{ and } \tau_3 = \tau_4 & : m_2^2 \delta(\tau_1 - \tau_2)\delta(\tau_3 - \tau_4) \\
\tau_1 = \tau_3 \text{ and } \tau_2 = \tau_4 & : m_2^2 \delta(\tau_1 - \tau_3)\delta(\tau_2 - \tau_4) \\
\tau_1 = \tau_4 \text{ and } \tau_2 = \tau_3 & : m_2^2 \delta(\tau_1 - \tau_4)\delta(\tau_2 - \tau_3) \\
0 & \text{otherwise}
\end{array}
\right.
\end{array}
\qquad \text{(A5.1.5)}
$$

In Section 5.3.3 we have to evaluate a case where we know that one pair of delays cannot be equal. Note that in such a case we have to combine from alternatives **VI–VIII**. For example if $\tau_2 \neq \tau_3$, we have two possibilities for pair forming:

(a) $\tau_1 = \tau_2$ and $\tau_3 = \tau_4$ in which pair τ_1, τ_2 is independent from pair τ_3, τ_4
(b) $\tau_1 = \tau_3$ and $\tau_2 = \tau_4$ in which pair τ_1, τ_3 is independent from pair τ_2, τ_4.

Now we can write the Expectation for $\tau_2 \neq \tau_3$ as the sum of (a) and (b):

$$E\{x(t - \tau_1)x(t - \tau_2)x(t - \tau_3)x(t - \tau_4)\}_{\tau_2 \neq \tau_3}$$

$$= \overbrace{E\{x(t - \tau_1)^2 x(t - \tau_3)^2\}}^{\text{case } a} + \overbrace{E\{x(t - \tau_1)^2 x(t - \tau_2)^2\}}^{\text{case } b}$$

$$= \underbrace{E\{x(t - \tau_1)^2\}}_{m_2} \underbrace{E\{x(t - \tau_3)^2\}}_{m_2} + \underbrace{E\{x(t - \tau_1)^2\}}_{m_2} \underbrace{E\{x(t - \tau_2)^2\}}_{m_2}$$

$$= m_2^2 \delta(\tau_1 - \tau_2)\delta(\tau_3 - \tau_4) + m_2^2 \delta(\tau_1 - \tau_3)\delta(\tau_2 - \tau_4) \qquad \text{(A5.1.6)}$$

Appendix 5.2

Creating Impulse Trains Following a Poisson Process

For the generation of a series of random numbers following a Gaussian or a uniform distribution, we use MATLAB commands randn and rand, respectively. A standard MATLAB command for generating a series of intervals according to a Poisson process does not exist. Therefore, we will apply a Monte Carlo technique to create such an impulse train according to a Poisson process. Our target is to follow a Poisson process characterized by probability density function (PDF) $\rho e^{-\rho x}$ (see Chapter 14 in van Drongelen, 2007). This works as follows. First we generate pairs of independent random numbers x,y with the MATLAB rand command. Because the rand command generates numbers between 0 and 1, x is multiplied with the maximal epoch value we want to consider, in order to rescale it between 0 and the maximum interval. Second, for each trial we compute $p = \rho e^{-\rho x}$, which is the probability p for interval x to occur according to the Poisson process. So far we will generate intervals x where all intervals have an equal probability because the MATLAB rand command is uniformly distributed. The second random number y associated with the randomly generated interval will also be evenly distributed between 0 and 1. We now only include pairs x,y in our series if $y < p$ and discard all others (Fig. A5.2.1); by following this procedure, the accepted intervals x obey the Poisson process because the probability that they are retained is proportional with $\rho e^{-\rho x}$, which is the desired probability. This procedure can, of course, be used for other distributions as well; it is known as the accept—reject algorithm.

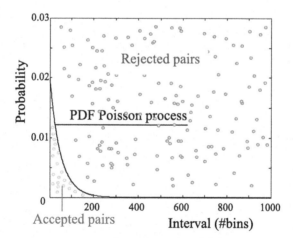

Figure A5.2.1 The Poisson process PDF can be used to create series of intervals obeying a Poisson process. Pairs of random uniformly distributed numbers x,y are generated: x is scaled between 0 and the maximum epoch length (1000 in this example) and y between 0 and 1. Each pair is then plotted in the X—Y plane. If $y < \rho e^{-\rho x}$ the point is accepted (green); otherwise it is rejected (red). If sufficient numbers are evaluated, the result is that epochs are retained according to the PDF describing the Poisson process.

The following MATLAB snippet of the function **Poisson.m** *shows an implementation of the procedure to generate a series of intervals following a Poisson process. This function is applied in* **pr5_1.m**. *Note that this routine also avoids intervals that are smaller than one bin because we do not allow for superimposed impulses.*

```
i=1;
while (i < len)
        x=rand;y=rand;        % two random numbers scaled 0-1
        x=x*epoch;            % the interval x is scaled 0-epoch
        p=rate*exp(-rate*x);  % the probability associated with the interval
                              % using the second random number using the
                              % Poisson process PDF
        if (y < p);           % Is the probability below the random # ?
        if x > 1;             % Avoid intervals that are too small ( < 1 bin)
        series(i)=x;          % else the interval is included
        i=i+1;
        end;
        end;
end;
```

6 Decomposition of Multichannel Data

6.1 Introduction

In the previous chapters we mainly focused on the analysis of single-input/single-output systems, single-channel data, or single images. Even when we worked with images, we worked with one row or column of pixels at a time. At most, we considered pairs of signals when we determined cross-correlation or coherence, or when we looked into input–output relationships. Although these techniques form a basis for analysis in neuroscience research, current studies usually collect multiple channels and/or movies of neural activity.

Examples of commonly encountered multichannel data sets are electroencephalograms (EEG), electrocorticograms (ECoG), recordings with multi-electrode arrays, a sequence of functional magnetic resonance images (fMRI), or movies made from neural tissue with voltage-sensitive or calcium indicator dyes. In these examples we deal not just with two or three simultaneously recorded signals, but with potentially overwhelming numbers of channels consisting of both spatial and temporal components. In the ECoG, each channel is at a certain location recording signals evolving over time; in both an fMRI sequence and a movie, the neural signals are represented by the intensity of each pixel as a function of time. Suppose we digitized an fMRI set with 128 samples in time where each image is 128×128 pixels, we would now have a huge data set consisting of $128^3 = 2,097,152$ points. Say we sample ECoG at a rate of 400 samples per channel per second and we record 128 channels for 60 s, resulting in a 1-min data set of $128 \times 400 \times 60 = 3,072,000$ data points. Examples of a small part of a 128-channel ECoG recording and a 21-channel EEG are shown in Fig. 6.1.

Typical goals for multichannel data processing are data reduction, decomposition, or investigation of the causal structure within the data. In the case of data reduction, we attempt to find the signal and noise components, and in the case of decomposition, our goal is to find the basic signal components that are mixed into the multichannel data set. Of course, both of these approaches are related. Suppose we have a measurement of brain activity during a task, and the activity associated with the task is signal while the remaining activity can be considered noise. If we can decompose our measured brain activity into these basic components, we have effectively used decomposition as a tool for data reduction.

Signal Processing for Neuroscientists, A Companion Volume. DOI: 10.1016/B978-0-12-384915-1.00006-1

(A)

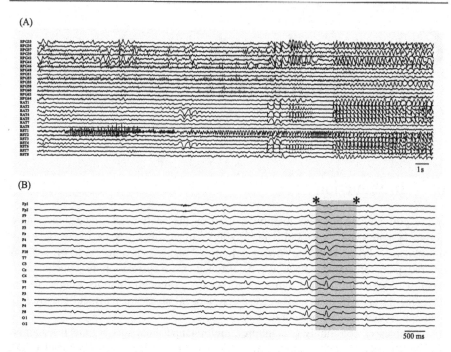

(B)

Figure 6.1 (A) Part of an intracranial 128-channel recording from a patient with epilepsy. Only a limited number of channels are shown: 15 channels of a right parietal grid (RPG), 7 channels from a right anterior temporal strip (RAT), and 7 from a right superior temporal strip (RST). The activity associated with a seizure onset starts with rapid oscillations in channel RST1 and eventually results in a spread of oscillatory activity in almost all channels. (B) A 21-channel scalp EEG recording from a different patient with epilepsy. The large-amplitude waveforms in this recording are epileptic spikes. The interval in between the asterisks (highlighted in yellow) is used for further analysis in Fig. 6.15. *(Color in electronic version.)*

   Although we will discuss multichannel analysis of data sets with large numbers of signals, we will introduce examples with strongly reduced data sets of two or three channels. The idea is to illustrate principles that are valid in high-dimensional data space with 2D or 3D examples. Throughout this chapter, we will demonstrate principles rather than formally prove them. First, we will show the general principle of mixing and unmixing of signals (Section 6.2), and then we will go into the details of specific strategies: principal component analysis (PCA; Section 6.3) and independent component analysis (ICA; Section 6.4). If you are interested in proofs, see texts on this topic (e.g., Bell and Sejnowski, 1995; Stone, 2004), on linear algebra (e.g., Jordan and Smith, 1997; Lay, 1997), and on information theory (e.g., Cover and Thomas, 1991; Shannon and Weaver, 1949).

6.2 Mixing and Unmixing of Signals

The underlying model of decomposition is that the signals of interest are mixtures of sources. For instance, our ear detects sounds from different sources

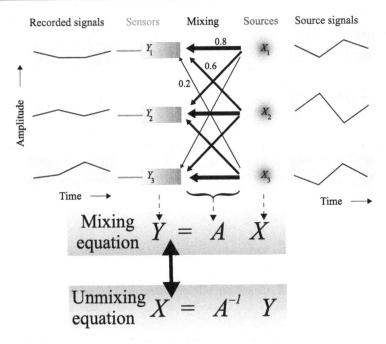

Figure 6.2 A mix of signals originating from three sources X (X_1-X_3) is recorded by three sensors Y (Y_1-Y_3). The signal from source to sensor is attenuated with distance. The amount of attenuation is symbolized by the width of the arrows (there are only three widths in this example: 0.8, 0.6, or 0.2). The mixing process can be represented by the matrix multiplication $Y = AX$, in which A is the mixing matrix. The unmixing process can be represented by $X = A^{-1}Y$, in which the unmixing matrix A^{-1} is the inverse of A. (*Color in electronic version.*)

simultaneously, such as someone talking to us, the noise of a fan, and music from a radio. Another example is an EEG electrode that picks up electrical activity from several parts of the brain. In other words, a set of measured signals Y_n consists of channels that are mixtures from a number of sources X_n. For now we assume that the number of sources is equal to or smaller than the number of signals we record. In this case the problem of mixing and unmixing can be defined mathematically. Let us consider a concrete example where we have three sources X_1-X_3 and three sensors Y_1-Y_3 (Fig. 6.2). The sensors pick up the signal from each source, but the signal is attenuated when traveling from source to sensor and the attenuation is proportional with distance. The level of attenuation in Fig. 6.2 is indicated by the width of the arrows: the signal from the source closest to the sensor attenuates least, only by a factor of 0.8. The signals from the other sources attenuate more, depending on distance, by a factor of 0.6 or a factor of 0.2 respectively. If we look at sensor Y_1 we can see that it will pick up 0.8 × the signal from source X_1 plus

$0.6 \times$ the signal from source X_2 plus $0.2 \times$ the signal from source X_3. Similar rules can be found for the other sensors Y_2 and Y_3. The measurements at the three sensors in Fig. 6.2 can be represented by a linear system of three equations:

$$Y_1 = 0.8X_1 + 0.6X_2 + 0.2X_3$$
$$Y_2 = 0.6X_1 + 0.8X_2 + 0.6X_3$$
$$Y_3 = 0.2X_1 + 0.6X_2 + 0.8X_3$$

Because this system has three equations with three unknowns $X_1–X_3$ (note that $Y_1–Y_3$ are known because they are measured), we can solve the system of equations above for the source signals $X_1–X_3$. We can put the equations in matrix form:

$$Y = AX \tag{6.1a}$$

in which,

$$Y = \begin{bmatrix} Y_1 \\ Y_2 \\ Y_3 \end{bmatrix}, \quad X = \begin{bmatrix} X_1 \\ X_2 \\ X_3 \end{bmatrix}, \quad \text{and the mixing matrix } A = \begin{bmatrix} 0.8 & 0.6 & 0.2 \\ 0.6 & 0.8 & 0.6 \\ 0.2 & 0.6 & 0.8 \end{bmatrix}$$

represents the attenuation coefficients for the setup in Fig. 6.2. The fact that $A(i,j) = A(j,i)$ (i.e., the mixing matrix A is symmetric) is due to the symmetric setup of the example in Fig. 6.2; this is not necessarily the case for a generic mixing matrix. Now suppose that we have recorded a time series of four samples from the three sensors and we want to know the signals from the individual sources. Because we know that $Y = AX$ we can compute X by $X = A^{-1}Y$, where A^{-1} is the inverse of A (here we assume that the inverse of A exists). If we take an example where the sources emit the following signals at times $t_1–t_4$ (indicated as source signals in the right panel in Fig. 6.2):

	t_1	t_2	t_3	t_4
X_1:	24.1667	1.6667	33.3333	17.5000
X_2:	-32.5000	0.0000	-55.0000	-22.5000
X_3:	20.8333	3.3333	41.6667	17.5000

Our sensors will pick up:

$$Y = AX = \begin{bmatrix} 0.8 & 0.6 & 0.2 \\ 0.6 & 0.8 & 0.6 \\ 0.2 & 0.6 & 0.8 \end{bmatrix} \begin{bmatrix} 24.1667 & 1.6667 & 33.3333 & 17.5000 \\ -32.5000 & 0.0000 & -55.0000 & -22.5000 \\ 20.8333 & 3.3333 & 41.6667 & 17.5000 \end{bmatrix}$$

$$= \begin{bmatrix} 4.0 & 2.0 & 2.0 & 4.0 \\ 1.0 & 3.0 & 1.0 & 3.0 \\ 2.0 & 3.0 & 7.0 & 4.0 \end{bmatrix}$$

So our measurement (indicated as recorded signals in the left panel in Fig. 6.2) will be:

$$
\begin{array}{ccccc}
 & t_1 & t_2 & t_3 & t_4 \\
Y_1 : & 4 & 2 & 2 & 4 \\
Y_2 : & 1 & 3 & 1 & 3 \\
Y_3 : & 2 & 3 & 7 & 4
\end{array}
$$

Because we know mixing matrix A, we compute its inverse (if you want to check this example in MATLAB, the inverse of a matrix can be obtained with the inv command) so that we can estimate the source activity $\hat{X}$ from the measurements:

$$\hat{X} = A^{-1} Y \tag{6.1b}$$

That is:

$$
\underbrace{\begin{bmatrix}
24.1667 & 1.6667 & 33.3333 & 17.5000 \\
-32.5000 & -0.0000 & -55.0000 & -22.5000 \\
20.8333 & 3.3333 & 41.6667 & 17.5000
\end{bmatrix}}_{\hat{X}} =
$$

$$
\underbrace{\begin{bmatrix}
5.8333 & -7.5000 & 4.1667 \\
-7.5000 & 12.5000 & -7.5000 \\
4.1667 & -7.5000 & 5.8333
\end{bmatrix}}_{A^{-1}}
\underbrace{\begin{bmatrix}
4.0 & 2.0 & 2.0 & 4.0 \\
1.0 & 3.0 & 1.0 & 3.0 \\
2.0 & 3.0 & 7.0 & 4.0
\end{bmatrix}}_{Y}
$$

As you can see, our estimate $\hat{X}$ for X is perfect (except for any precision errors due to computation). Although this example clarifies the mixing and unmixing process, it is not very helpful in practical applications (even if we ignore the effects of noise that would be present in any real recording) because the mixing matrix is unknown and/or the number of sources outnumbers the number of sensors. In the remainder of this chapter we will focus on what one can do if the mixing matrix is unknown. In this case, we want to separate the sources while we are "blind" with respect to the mixing process; therefore, these procedures are called blind source separation (BSS). We will specifically focus on two of these techniques: PCA and ICA.

6.3   Principal Component Analysis

In this section we introduce the concept of decomposing multichannel data into its principal components. With PCA of a multidimensional measurement, one can find the directions of maximal and minimal variance in the multidimensional measurement space. We will see that these directions are orthogonal, indicating that the components extracted with PCA are uncorrelated. We will introduce the technique by analyzing a concrete 3D example of four measurements S_1–S_4, each observation S_n having three values or signals s_1, s_2, and s_3 (one for each of the three dimensions):

$$S_1 = \begin{bmatrix} 4 \\ 1 \\ 2 \end{bmatrix} \quad S_2 = \begin{bmatrix} 2 \\ 3 \\ 3 \end{bmatrix} \quad S_3 = \begin{bmatrix} 2 \\ 1 \\ 7 \end{bmatrix} \quad S_4 = \begin{bmatrix} 4 \\ 3 \\ 4 \end{bmatrix} \tag{6.2a}$$

The mean vector of these four observations, M, contains the mean for each of the three signals m_1, m_2, and m_3:

$$M = \frac{1}{4}\{S_1 + S_2 + S_3 + S_4\} = \frac{1}{4}\left\{ \begin{bmatrix} 4 \\ 1 \\ 2 \end{bmatrix} + \begin{bmatrix} 2 \\ 3 \\ 3 \end{bmatrix} + \begin{bmatrix} 2 \\ 1 \\ 7 \end{bmatrix} + \begin{bmatrix} 4 \\ 3 \\ 4 \end{bmatrix} \right\} = \begin{bmatrix} m_1 \\ m_2 \\ m_3 \end{bmatrix} = \begin{bmatrix} 3 \\ 2 \\ 4 \end{bmatrix} \tag{6.2b}$$

A 3D plot of the observations and their mean is shown in Fig. 6.3A. If we now demean our four observations—that is, we subtract M from S_1 to S_4 (as we generally do with our signals before processing them)—and we group the demeaned observation in matrix B, we have:

$$B = \begin{bmatrix} 4-3 & 2-3 & 2-3 & 4-3 \\ 1-2 & 3-2 & 1-2 & 3-2 \\ 2-4 & 3-4 & 7-4 & 4-4 \end{bmatrix} = \begin{bmatrix} 1 & -1 & -1 & 1 \\ -1 & 1 & -1 & 1 \\ -2 & -1 & 3 & 0 \end{bmatrix} \tag{6.3}$$

In statistics, a data set from multichannel observations such as the concatenated matrix $S = \begin{bmatrix} S_1 & S_2 & S_3 & S_4 \end{bmatrix}$ or matrix B is called multivariate data. A scatter plot of the demeaned observations is shown in Fig. 6.3B. Note that the new mean value is now at the origin, and so we have in effect translated the axes of our coordinate system. From B, we can compute the covariance matrix C. Since we have three variables (s_1, s_2, s_3) in each observation, the covariance matrix is 3×3. If we have N observations, each entry in the matrix can be computed by $C(i,j) = 1/(N-1) \sum_{n=1}^{N} (s_i - m_i)_n (s_j - m_j)_n$. In this example, C is a 3×3 matrix, i and j range from 1 to 3, and N is the number of observations, in this example $N = 4$ (since we have observations S_1–S_4). In matrix notation this notation can be simplified to:

$$C = \frac{1}{N-1} BB^T = \frac{1}{3} \begin{bmatrix} 1 & -1 & -1 & 1 \\ -1 & 1 & -1 & 1 \\ -2 & -1 & 3 & 0 \end{bmatrix} \begin{bmatrix} 1 & -1 & -2 \\ -1 & 1 & -1 \\ -1 & -1 & 3 \\ 1 & 1 & 0 \end{bmatrix}$$

$$= \begin{bmatrix} 1.33 & 0 & -1.33 \\ 0 & 1.33 & -0.67 \\ -1.33 & -0.67 & 4.67 \end{bmatrix} \tag{6.4}$$

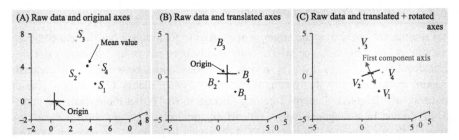

Figure 6.3 (A) A 3D plot of four observation vectors S_1-S_4 (Equation (6.2a)) and their mean value M (Equation (6.2b)). (B) The same points, now indicated as B_1-B_4 because they are plotted against axes that are translated so that the mean M becomes the new origin. (C) Finally we plot the same points (now indicated as V_1-V_4) against axes that are also rotated to reflect the directions of the three principal components. The first principal component is indicated by the double arrow (red). This illustration was made with MATLAB script Pr6_1.m (available on http://www.elsevierdirect.com/companions/9780123849151); the numerical values can be found in Table 6.1. (*Color in electronic version.*)

The superscript "T" indicates the transpose of matrix B (in the transpose rows and columns are interchanged such that $B(i,j) \rightarrow B(j,i)$). Each value in the diagonal of C represents the variance of the b_1, b_2, and b_3 values of observation vectors B_n. So the sum of the diagonal elements, the trace of C written as tr(C), is the total variance. Each off-diagonal element is a covariance value—for example, $C(2,3)$ is the covariance between the b_2 and b_3 coordinates. Of course, $C(3,2)$ is the same value because it is the covariance between the b_3 and b_2 coordinates. Therefore a covariance matrix is always a **symmetric** matrix (see the example in Equation (6.4)). A more formal way to establish symmetry for covariance matrices is to show that interchanging the rows and columns (transposition) of covariance matrix C results in the same matrix: that is, $C = C^T$. From Equation (6.4) we can establish that C is proportional with BB^T (by a factor of $1/(N-1)$) and the transposing operation on C can be represented by $(BB^T)^T = B^{TT}B^T$ (recall that the multiplication order of matrices switches when taking their transpose). Because the transpose of a transposed matrix is the original matrix again, we may simplify this outcome $B^{TT}B^T = BB^T$, which shows that $(BB^T)^T = BB^T$—that is, the transpose of BB^T is BB^T again.

If $C(i,j)$ for $i \neq j$ is zero, there is no covariance or correlation between b_i and b_j. It may be clear that analysis of multivariate data is simpler when all signals are uncorrelated—that is, a covariance matrix that is **diagonal**, which means that all off-diagonal elements are zero. This is exactly the goal of the decomposition with PCA.

Note: Correlation (ρ_{xy}) between two variables x and y is a normalized version of the covariance (Cov(x,y)) between x and y—that is, $\rho_{xy} = \text{Cov}(x,y)/\sigma_x\sigma_y$— with standard deviations σ_x and σ_y for x and y, respectively. The effect of this normalization is that the correlation coefficient ρ_{xy} is scaled between -1 and 1.

6.3.1 Finding Principal Components

To summarize the above, the strategy of PCA is to manipulate our demeaned observations B_n (b_1, b_2, b_3)$_n$ for which correlations between b_i and b_j may exist into transformed data V_n (v_1, v_2, v_3)$_n$ such that all correlations between v_i and v_j vanish. Again, mathematically this means that the covariance matrix C of B may contain nonzero off-diagonal elements (see, e.g., Equation (6.4)), but the covariance matrix Σ of V must be a diagonal matrix (all off-diagonal elements are zero). Let us continue with our example and use the PCA approach to find the components. We first introduce and apply the method; later we justify the procedure in the context of the above strategy.

Continuing the numerical example above, we will show that the 3 × 3 covariance matrix C in Equation (6.4) can be diagonalized by applying a linear transformation. To accomplish this, we first define a 3 × 3 matrix of orthogonal column vectors $U = \begin{bmatrix} U_1 & U_2 & U_3 \end{bmatrix}$ and a 3 × 3 diagonal matrix Σ with diagonal entries $\lambda_1 - \lambda_3$, and group our demeaned observations B_n in matrix B (Equation (6.3)). We can compute:

$$CU = \begin{bmatrix} CU_1 & CU_2 & CU_3 \end{bmatrix} \tag{6.5a}$$

and

$$U\Sigma = \begin{bmatrix} U_1 & U_2 & U_3 \end{bmatrix} \begin{bmatrix} \lambda_1 & 0 & 0 \\ 0 & \lambda_2 & 0 \\ 0 & 0 & \lambda_3 \end{bmatrix} = \begin{bmatrix} \lambda_1 U_1 & \lambda_2 U_2 & \lambda_3 U_3 \end{bmatrix} \tag{6.5b}$$

Note that Σ is a diagonal matrix. Now let us assume that our covariance matrix C is diagonalizable such that:

$$C = U\Sigma U^{-1} \quad \text{and} \quad \Sigma = U^{-1}CU \tag{6.5c}$$

(note that we also assumed that U is invertible). If we right-multiply the first expression in Equation (6.5c) by U we get:

$$CU = U\Sigma \tag{6.5d}$$

This result indicates that if C is diagonalizable, then the expressions in Equations (6.5a) and (6.5b) must be equal. If we equate the individual columns in the matrices in Equation (6.5d), we get:

$$CU_1 = \lambda_1 U_1, \quad CU_2 = \lambda_2 U_2, \quad \text{and} \quad CU_3 = \lambda_3 U_3 \tag{6.5e}$$

The result in Equation (6.5e) shows that $\lambda_1 - \lambda_3$ and $U_1 - U_3$ must be the eigenvalues and corresponding eigenvectors of the covariance matrix C. See Appendix 6.1 if you need to review the concept of eigenvalues and eigenvectors; if you need more than a quick review, see a text on linear algebra such as the first part of Jordan and Smith (1997) or Lay (1997).

Because C is a symmetric matrix, its eigenvectors are orthogonal vectors. We can show this property of symmetric matrices by considering a simple 2D case where we have two distinct eigenvalues (λ_1 and λ_2) with two corresponding eigenvectors (U_1 and U_2). To show that these vectors are orthogonal, we show that their scalar product equals zero.

Note: Recall that the inner product (also called scalar product or dot product) of two vectors $\vec{a}$ and $\vec{b}$ is given by $ab \cos \phi$, where a and b are the lengths of the vectors and ϕ is the angle between them. If the vectors are orthogonal, ϕ equals 90° and the outcome of the dot product is zero.

We can show that the dot product $U_1 \cdot U_2 = 0$ by computing the following expression:

$$\lambda_1 U_1 \cdot U_2 = (\lambda_1 U_1)^{\mathrm{T}} U_2 = (CU_1)^{\mathrm{T}} U_2 = U_1^{\mathrm{T}} C^{\mathrm{T}} U_2 \tag{6.6a}$$

Here we changed the vector dot product notation into vector notation $U_1 \cdot U_2 = U_1^{\mathrm{T}} U_2$ (note the presence of the dot in the far-left expression), and we used the definition of the eigenvalue/eigenvector of C: $\lambda_1 U_1 = CU_1$ (Appendix 6.1). We know that C is a covariance matrix that must be symmetric; therefore, $C = C^{\mathrm{T}}$. Using this property for symmetric matrices, we get:

$$U_1^{\mathrm{T}} C^{\mathrm{T}} U_2 = U_1^{\mathrm{T}} (CU_2) = U_1^{\mathrm{T}} (\lambda_2 U_2) = \lambda_2 U_1^{\mathrm{T}} U_2 = \lambda_2 U_1 \cdot U_2 \tag{6.6b}$$

Note the dot in the last expression. Combining Equations (6.6a) and (6.6b), we may conclude that for the symmetric covariance matrix:

$$\lambda_1 U_1 \cdot U_2 = \lambda_2 U_1 \cdot U_2 \rightarrow (\lambda_1 - \lambda_2) U_1 \cdot U_2 = 0 \tag{6.6c}$$

Because we deal with two distinct eigenvalues, we know that $(\lambda_1 - \lambda_2) \neq 0$ and therefore the scalar product $U_1 \cdot U_2 = 0$, indicating that the angle between the two eigenvectors of a symmetric matrix must be 90°. Thus the two vectors are orthogonal (perpendicular):

$$U_1 \perp U_2 \tag{6.6d}$$

So if we need an orthogonal matrix, we can use the orthogonal eigenvectors of the covariance matrix to create the matrix U to transform the observed demeaned data.

Let us apply the results from the above paragraphs to our numerical example (Equations (6.2)–(6.4)). First we must find a 3 × 3 matrix of orthogonal eigenvectors vectors $U = \begin{bmatrix} U_1 & U_2 & U_3 \end{bmatrix}$ to transform the demeaned data—that is, $B = UV$.

Matrix V contains the transformed vectors V_1-V_4. This means that for each demeaned observation B_n we want to identify an orthogonal change of variable V_n such that:

$$B_n = UV_n \rightarrow \begin{bmatrix} b_1 \\ b_2 \\ b_3 \end{bmatrix}_n = \begin{bmatrix} U_1 & U_2 & U_3 \end{bmatrix} \begin{bmatrix} v_1 \\ v_2 \\ v_3 \end{bmatrix}_n \qquad (6.7)$$

Recall that in the above U_1-U_3 are column vectors so that U is a 3×3 matrix $u_{i,j}$, that is, $b_1 = u_{1,1} \times v_1 + u_{1,2} \times v_2 + u_{1,3} \times v_3$, etc. Assuming again that U is invertible, we can write the relationship in Equation (6.7) as $V_n = U^{-1}B_n$. Since U is an orthogonal matrix, its inverse is equal to its transpose (see a linear algebra text such as Lay, 1997, if you need to review this), so we may write $U^{-1}B_n = U^{T}B_n$. Recalling how we computed the covariance matrix C from B and its transpose (Equation (6.4)), we can now find the covariance matrix Σ for V:

$$\begin{aligned} \Sigma &= \frac{1}{N-1}VV^{T} = \frac{1}{N-1}(U^{T}B)(U^{T}B)^{T} \quad \text{since} \quad V = U^{T}B \\ &= \frac{1}{N-1}U^{T}BB^{T}U \qquad\qquad\qquad \text{since } (U^{T}B)^{T} = B^{T}U \\ &= U^{T}\underbrace{\frac{1}{N-1}BB^{T}}_{C:\ \text{Equation (6.4)}}U = U^{T}CU \end{aligned} \qquad (6.8a)$$

So the orthogonal matrix U can relate C to Σ:

$$\Sigma = U^{T}CU = U^{-1}CU \qquad (6.8b)$$

In the above we used again $U^{-1} = U^{T}$ to obtain a result for Σ that is the same as the second expression in Equation (6.5c). Thus the covariance matrix for transformed observations V_n is the diagonal matrix Σ. Because the off-diagonal elements (the covariance values) of Σ are zero, v_1, v_2, and v_3 of the transformed observations are uncorrelated. The diagonal elements of Σ, eigenvalues $\lambda_1-\lambda_3$, are the variance values for the transformed observations v_1-v_3. Convention for PCA is that the eigenvalues and associated eigenvectors are sorted from high to low eigenvalues (variance).

6.3.2 A MATLAB Example

If we compute the eigenvalues and eigenvectors for covariance matrix C, we can transform our demeaned observations depicted in Fig. 6.3B. In MATLAB this can be easily accomplished with the eig command—that is, [UU,SIGMA] = eig(C). In our example, we obtain three eigenvectors that form a rotated set of axes relative to the translated axes in Fig. 6.3B. This is because the eigenvectors are orthogonal (i.e., perpendicular) (Equation (6.6d)). If we arrange the eigenvectors according to the magnitude of their associated eigenvalues (variance), we get the first, second, and third principal components (note then that the first principal component is along the direction of greatest variance). In Fig. 6.3C the first component is indicated by a double arrow (red) and the remaining two components by lines (black); in this example it is easy to see that the first component is in the direction of maximal variance. The covariance matrix C and its eigenvectors and eigenvalues (grouped in Σ and sorted for the eigenvalues in **descending** order) are:

$$C = \begin{bmatrix} 1.3333 & 0 & -1.3333 \\ 0 & 1.3333 & -0.6667 \\ -1.3333 & -0.6667 & 4.6667 \end{bmatrix}$$

(See also Equation (6.4))

$$U = \begin{bmatrix} -0.3192 & 0.4472 & 0.8355 \\ -0.1596 & -0.8944 & 0.4178 \\ 0.9342 & 0 & 0.3568 \end{bmatrix}$$

and

$$\Sigma = \begin{bmatrix} 5.2361 & 0 & 0 \\ 0 & 1.3333 & 0 \\ 0 & 0 & 0.7639 \end{bmatrix}$$

Note: If you do this example in MATLAB, the eig command sorts the eigenvalues from low to high. This is in **ascending** order, which is contrary to convention for PCA. Therefore, the order of the diagonal entries in SIGMA and Σ and the order of the associated eigenvectors (columns) in UU and U are reversed.

Suppose we want to find the coordinates of our observations S_1–S_4 on the translated-and-rotated set of axes (Fig. 6.3C)—in other words, the projections of the observations on the eigenvectors. Let us look into our numerical example how this can be accomplished by computing the projection of the first observation on the first principal component. First, we take point B_1 (corresponding to a demeaned version of the first observation S_1 in Equation (6.2a) and depicted in Fig. 6.3B), which is the first column of B in Equation (6.3):

$$B_1 = \begin{bmatrix} 1 \\ -1 \\ -2 \end{bmatrix}$$

Now let us take the first eigenvector, which is the first column of matrix U:

$$U_1 = \begin{bmatrix} -0.3192 \\ -0.1596 \\ 0.9342 \end{bmatrix}$$

The projection of the first point (black in Fig. 6.3) on this eigenvector can be determined by the scalar product of the two vectors:

$$B_1 \cdot U_1 = B_1^T U_1 = \begin{bmatrix} 1 & -1 & -2 \end{bmatrix} \begin{bmatrix} -0.3192 \\ -0.1596 \\ 0.9342 \end{bmatrix} = -2.0279$$

The above can easily be checked in MATLAB after running the example program Pr6_1. m (available on http://www.elsevierdirect.com/companions/9780123849151). Use B(:,1) and U(:,1) for B_1 and U_1 respectively; the scalar product can be computed with B (:,1)'*U(:,1) (note the ' for transposing B(:,1)). The outcome is -2.0279, the projection of the first point on the first eigenvector. The projection of the first point on the second and third eigenvectors will be scalar products $B_1 \cdot U_2$ and $B_1 \cdot U_3$ (note the dots). For the second point B_2 (red in Fig. 6.3) we can repeat the procedure: $B_2 \cdot U_1$, $B_2 \cdot U_2$, and $B_2 \cdot U_3$. The same, of course, is true for the third (blue, Fig. 6.3) and the fourth (green, Fig. 6.3) points. We can compute all the scalar products V at once with the matrix multiplication $B^T U$. This will generate the coordinates of all four points on the three eigenvectors. The results for our numerical example are summarized in Table 6.1.

Table 6.1 Principal Component Analysis: Numerical Example

$S =$	$[S_1$	S_2	S_3	$S_4]$	Original observations
s_1	4.0000	2.0000	2.0000	4.0000	Fig. 6.3A
s_2	1.0000	3.0000	1.0000	3.0000	
s_3	2.0000	3.0000	7.0000	4.0000	
$B =$	$[B_1$	B_2	B_3	$B_4]$	Demeaned observations
b_1	1.0000	-1.0000	-1.0000	1.0000	Fig. 6.3B
b_2	-1.0000	1.0000	-1.0000	1.0000	
b_3	-2.0000	-1.0000	3.0000	0	
$V =$	$[V_1$	V_2	V_3	$V_4]$	Projections on eigenvectors
v_1	$\mathbf{-2.0279}$	-0.7746	$\mathbf{3.2812}$	-0.4787	Fig. 6.3C
v_2	1.3416	-1.3416	0.4472	-0.4472	
v_3	-0.2959	-0.7746	-0.1829	1.2533	

Summary of PCA on four observations S_1-S_4. These data points are plotted in Fig. 6.3A. First the data are demeaned in B_1-B_4 so that a new set of axes with its origin in the point of gravity of all points is obtained (Fig. 6.3B). Finally the axes are rotated using the PCA (Fig. 6.3C). Note that the first component axis (double arrow, red in Fig. 6.3C) indicates the direction of largest variance, easily appreciated when looking at the position of the first (V_1, black) and third (V_3, blue) points in Fig. 6.3C. For clarity, these extreme values for the first component v_1 are indicated in bold in the table (v_1 in vectors V_1 and V_3). *(Color in electronic version.)*

Note: In some texts the projection on the first eigenvector (row v_1 in Table 6.1) is indicated as the first principal component, the projections v_2 on the second eigenvector is then the second principal component, etc. To summarize, depending on the text, the principal components can be the eigenvectors U_1-U_3 or the projections of the observations on these vectors v_1-v_3, and in some texts the term "principal component" is used for both.

The variance in each direction (i.e., for each component v_1, v_2, and v_3) is easily calculated in MATLAB after running the program Pr6_1.m with the std command: std(V').^2 . The outcome of this calculation is 5.2361, 1.3333, 0.7639; as expected, these values correspond to the eigenvalues in Σ. Because the origin of the axes in Fig. 6.3C is the same as in panel B, the mean of the components v_1-v_3 remains zero (mean(V')). Further we can test for zero covariance—that is, testing that the off-diagonal entries of the covariance matrix (1/3)*V*V' are indeed zero. The result is:

5.2361	0.0000	−0.0000
0.0000	1.3333	0.0000
−0.0000	0.0000	0.7639

The outcome is as expected: the diagonal elements are again the variances for v_1-v_3 and all covariance values are zero.

6.3.3 Singular Value Decomposition

A common technique to compute the eigenvalues and eigenvectors of the covariance matrix directly from the demeaned observations is singular value decomposition. This technique is based on the fact that any rectangular matrix, such as the demeaned observation matrix B, can be decomposed as:

$$B = U\Theta W^T \tag{6.9}$$

Note that this expression looks similar to the first expression in Equation (6.5c). In Equation (6.9) U and W are orthogonal matrices, and Θ is a matrix that includes a matrix Σ for which the diagonal entries are the so-called singular values σ_1, σ_2, σ_3, ..., σ_r. In our numerical example above where B is a 3×4 matrix (Equation (6.3)), U is a 3×3 matrix of eigenvectors, W is a 4×4 matrix of eigenvectors, and Θ is the same size as B, a 3×4 matrix in which the first 3×3 diagonal entries are the singular values $\sigma_1-\sigma_3$. In this example:

$$B = \begin{bmatrix} 1 & -1 & -1 & 1 \\ -1 & 1 & -1 & 1 \\ -2 & -1 & 3 & 0 \end{bmatrix} \quad U = \begin{bmatrix} -0.3192 & 0.4472 & 0.8355 \\ -0.1596 & -0.8944 & 0.4178 \\ 0.9342 & -0.0000 & 0.3568 \end{bmatrix}$$

$$\Theta = \begin{bmatrix} 3.9634 & 0 & 0 & 0 \\ 0 & 2.0000 & 0 & 0 \\ 0 & 0 & 1.5139 & 0 \end{bmatrix}$$

$$W = \begin{bmatrix} -0.5117 & 0.6708 & -0.1954 & 0.5000 \\ -0.1954 & -0.6708 & -0.5117 & 0.5000 \\ 0.8279 & 0.2236 & -0.1208 & 0.5000 \\ -0.1208 & -0.2236 & 0.8279 & 0.5000 \end{bmatrix}$$

While the eigenvectors (columns of U) we find here correspond with those found for the covariance matrix above, you may be surprised that the singular values in Θ do not correspond with those in Σ above. This is because (unlike the eigenvalues $\lambda_1 - \lambda_3$ of the covariance matrix C) the singular values σ_i are the standard deviations and not the variance. Furthermore, the singular values are based on BB^T while the eigenvalues λ_i are based on the normalized version: BB^T divided by $(N - 1)$. So if we compute $\Theta\Theta^T$ and divide by $N - 1 = 3$, we get the same values as the diagonal entries in Σ:

$$\Sigma = \frac{\Theta\Theta^T}{N - 1} = \frac{\Theta\Theta^T}{3} = \begin{bmatrix} 5.2361 & 0 & 0 \\ 0 & 1.3333 & 0 \\ 0 & 0 & 0.7639 \end{bmatrix}$$

This result is identical to the values we obtained for covariance matrix Σ we obtained earlier. If we use Equation (6.9) to compute BB^T:

$$BB^T = (U\Theta W^T)(U\Theta W^T)^T = (U\Theta W^T)(W\Theta^T U^T) = U\Theta \underbrace{W^T W}_{I} \Theta^T U^T$$

$$= U\underbrace{\Theta\Theta^T}_{\Sigma''} U^T = U\Sigma'' U^T \tag{6.10}$$

In the above we used $W^{TT} = W$. Since W is orthogonal, $W^T = W^{-1}$, so we may state that $W^T W = I$, where I is the identity matrix. Finally, because Θ's only non-zero entries are on the diagonal, we may state: $\Theta\Theta^T = (N - 1)\Sigma = \Sigma''$. Recalling that BB^T divided by $(N - 1)$ is the covariance matrix C, the outcome of Equation (6.10) is (with the exception of the normalization $1/(N - 1)$, reflected by the use of Σ'' instead of Σ) the same as the left expression in Equation (6.5c). Restating this here for convenience: $C = U \Sigma U^{-1}$ (recall that $U^T = U^{-1}$ because U is an orthogonal matrix).

We can use standard MATLAB functions to compute the eigenvalues and eigenvectors from the covariance matrix using the eig *command, or directly from the demeaned observations using singular value decomposition with the* svd *command. A part of* Pr6_1.m *(available on http://www.elsevierdirect.com/companions/9780123849151) shows the use of these commands.*

```
% Two Methods to Perform PCA using MATLAB standard functions eig and svd
% 1. Eigenvalues and Eigenvectors (eig) of Covariance Matrix C
% [=(1/(N-1)*B*B']
[ei_vectors1,ei_values1]=eig(C)
['NOTE that the eigenvalues above are sorted in ASCENDING order']

% 2. Singular Value Decomposition (svd) of DEMEANED Observation Matrix B
[ei_vectors2,singular_values,vv]=svd(B)
['NOTE that the eigenvalues above are sorted in DESCENDING order']
% IMPORTANT NOTE
% singular_values is the sqrt of the eigenvalues of the non-normalized
% covariance B*B' [i.e. sqrt(eig(B*B'))]
```

As a final note, you can see that the PCA would do a bad job distinguishing source signals from a mixture, since PCA separates components based purely on variance. In our computations above, our S matrix was the same as the measured signals Y in the example of Fig. 6.2 in Section 6.2. However, the temporal sequences in the decomposed results in V (Table 6.1) do not even come close to unmixing the source signals X in that example. In the following section we will see how PCA can be used with more success as a tool to separate signal(s) from noise.

6.3.4   Using PCA as a Filter

The PCA technique detects uncorrelated components with decreasing variance. One application that uses this property is to remove noisy components from mixtures of signals. The reasoning for this application is that signal components should display high variance, while the added noise components have smaller variance. Of course, the truth of this assumption depends on the type of signal and may not always be valid.

In the MATLAB script Pr6_2.m (available on http://www.elsevierdirect.com/companions/9780123849151) we explore this technique by purposely corrupting an image (Lena) with random noise to examine how well we can clean up the mess using PCA. Subsequently we use singular value decomposition to define the principal components. The program displays a series of 30 figures each with four panels: the original image, its noisy contaminated version, the image reflecting the nth principal component, and the image reflecting the sum of components 1 to n. It can be seen that the PCA cannot retrieve the original image, but it certainly can improve the noisy contaminated version. At some point (around component 10−15 in this example) the higher components do not seem to further improve image quality in the sum of the components 1 to n. This is due to the fact that the higher components indeed contain more of the noisy aspect of the corrupted image, thus decreasing corrupted image quality when added to the sum of components.

6.4 Independent Component Analysis

In the previous section we introduced PCA, a technique to decompose multichannel data into uncorrelated components. If we use PCA to decompose x and y into variables a and b, we have shown that the covariance between decomposed variables a and b is zero and the covariance matrix of these transformed observations is entirely determined by variance. ICA moves beyond the constraint of decorrelation and looks for components that are statistically independent. When two signals a and b are statistically independent, they are each drawn from an independent probability density function (pdf), and the joint pdf of $[a\ b]$ is simply the product of the individual pdfs:

$$p_{ab}([ab]) = p_a(a)p_b(b) \tag{6.11}$$

The joint and individual pdfs are symbolized by p_{ab}, p_a, and p_b, respectively. Suppose we have two processes a and b, with probability distributions $p_a = \begin{bmatrix} 0.1 & 0.2 & 0.2 & 0.4 & 0.1 \end{bmatrix}$ and $p_b = \begin{bmatrix} 0.1 & 0.3 & 0.3 & 0.2 & 0.1 \end{bmatrix}$; then, if they are independent, we may use Equation (6.11) to compute the joint probability p_{ab} (Fig. 6.4). Note that the probability functions in Fig. 6.4 all add up to 1 ($\sum_{i=1}^{5} p_{a_i} = 1$, $\sum_{i=1}^{5} p_{b_i} = 1$, and $\sum_{i=1}^{5} \sum_{j=1}^{5} p_{a_i b_j} = 1$). Statistical independence between a and b means that all the moments and central moments (moments about the mean) of the distributions for a and b must also be independent:

$$E\{a^p b^q\} = E\{a^p\}E\{b^q\} \tag{6.12a}$$

Here $E\{\dots\}$ denotes the Expectation (see section 3.2 in van Drongelen, 2007, if you need to refresh your knowledge about Expectation). If a and b are demeaned, the first central moment (exponents $p = 1$ and $q = 1$ in Equation (6.12a)) of the

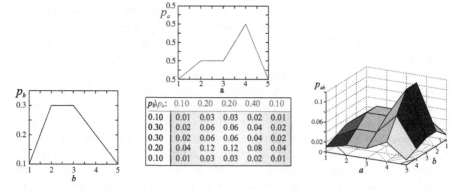

Figure 6.4 An example of a 2D joint probability density function p_{ab} (in the green panel) and its marginal distributions p_a and p_b. The graphs show the individual, marginal distributions of a (top graph) and b (left graph). In the 3D graph on the right, the joint probability is plotted on the vertical axis against the variables a and b. (*Color in electronic version.*)

joint pdf $E\{a\ b\}$ is the covariance. If a and b are uncorrelated (as in the decomposed result from PCA), we have:

$$E\{ab\} = \underbrace{E\{a\}}_{0}\ \underbrace{E\{b\}}_{0} = 0 \tag{6.12b}$$

You can see that demanding that a and b are uncorrelated (Equation (6.12b)) is not as strong a condition as asking for statistical independence of a and b (Equation (6.12a)). There is an exception when PCA does generate statistically independent components: this is the case when the extracted signals are normally distributed. Normally distributed signals are determined by their first two moments; once these are known, all higher-order moments are determined. **To summarize: two signals that are statistically independent are also uncorrelated, but uncorrelated signals are not statistically independent except when the signals are normally distributed.**

In real cases, signal mixtures tend to be normally distributed due to the central limit theorem. To put it informally, the central limit theorem states that the sum or a mixture (=weighted sum) of multiple variables tends to be normally distributed even when individual components are not drawn from a normal distribution. An example of this theorem at work is shown in Fig. 6.5. In this example we study a mixture of variables that are each uniformly distributed (Fig. 6.5A). Interestingly, the mixture of only five such variables already shows a tendency toward a normal distribution (Fig. 6.5B). This shows that the practical application of PCA for extracting statistically independent components will be fairly limited because (unlike signal mixtures) it is less likely that individual signal components are normally distributed. Because ICA demands statistical independence of the individual components, ICA is much better at extracting components that are not normally distributed. To visualize the difference in component distributions, we can look at the distribution of observations from a uniform distribution (Fig. 6.5A), where we observe that the points are scattered more or less evenly over a line in the 1D case (Fig. 6.5C) or a plane in the 2D case (Fig. 6.5E). On the other hand, normally distributed (Gaussian) mixtures are concentrated around the mean value of the distribution (Fig. 6.5B, D, and F).

6.4.1 Entropy of Sources and Mixtures

Recall that the entropy $S(X)$ of a random variable X (see van Drongelen, 2007, chapter 14, section 14.3) depends on the probability distribution of X. It can be defined as the sum (in the case of a discrete variable) or the integral (in case of a continuous variable) of the product $p(x)\log_2 1/(p(x)) = -p(x)\log_2 p(x)$ over all x:

$$S(X) = -\sum_{\text{All } x} p(x)\log_2 p(x) \tag{6.13}$$

In this case we defined S for a discrete variable and we use $\log_2$, a base 2 logarithm, so that S is in units of bits.

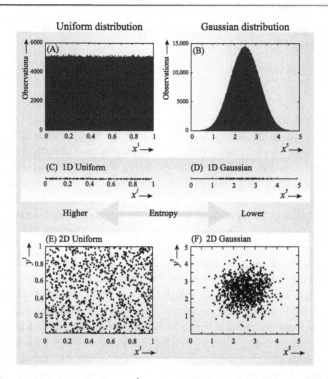

Figure 6.5 (A) Histogram of a variable x^1 that is uniformly distributed between 0 and 1. (B) The sum of only five of these uniformly distributed variables x^5 tends to be almost normally distributed. (A) and (B) were made with script Pr6_3.m (available on http://www.elsevierdirect. com/companions/9780123849151). A series of one-dimensional observations from uniform (x^1) and (almost) Gaussian (x^5) distributions are shown in (C) and (D), respectively. The scatter plots in (E) and (F) are examples of a series of 2D observations: two variables $x^1 y^1$ for the uniform case, and two variables $x^5 y^5$ for the Gaussian one. As expected, the uniform distribution results in a scatter of points throughout the plane, whereas the Gaussian case shows a concentration of points around a center (the mean). Consequently, the entropy of the uniformly distributed points is higher than the entropy for the Gaussian distributed observations. *(Color in electronic version.)*

Let us consider a very simple case, a coin toss. If we have the usual situation, we have probability $p = \frac{1}{2}$ for both outcomes heads and tails (scenario III, Fig. 6.6A), and the entropy according to Equation (6.13) is:

$$S(X) = -\left[\frac{1}{2}\log_2\frac{1}{2} + \frac{1}{2}\log_2\frac{1}{2}\right] = 1 \text{ bit}$$

This is a reasonable result, because we have an outcome that fits in a single bit: either heads (1) or tails (0). Now suppose we have a "faulty" (deterministic) coin that always lands on one side, either heads or tails. In these scenarios (I and V, Fig. 6.6A) we have $p = 0$ for one outcome and $p = 1$ for the other; now the entropy is:

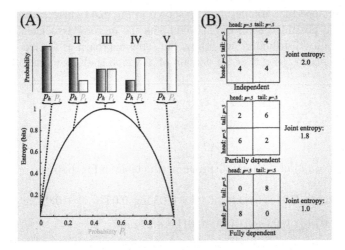

Figure 6.6 Statistics of a coin toss and entropy. (A) Five scenarios of probability distributions of heads (p_h) and tails (p_t). The graph depicts that each scenario is associated with a specific entropy value. (B) Statistics of coin tosses. The numbers in the tables show the idealized outcomes for 16 tosses. Two coins are used for each observation and in the upper diagram the outcomes of each toss is completely independent. In the two lower diagrams, there (magically) is some dependence between the two tosses in each observation—there is either a full dependence (bottom diagram; the pair of outcomes in each observation are identical) or a partial dependence (middle diagram). *(Color in electronic version.)*

$$S(X) = -\left[0 \log_2 0 + 1 \log_2 1\right] = 0 \text{ bit}$$

Note that we use $0 \log_2 0 = 0$. This outcome of zero bit also seems reasonable since there is no surprise (information) with each outcome: it will always be heads in one scenario and always tails in the other. If our coin is biased and we get heads or tails in 75% of the cases (scenarios II and IV in Fig. 6.6A), we have probabilities $p = 0.75$ and $p = 0.25$ and the associated entropy is:

$$S(X) = -\left[\frac{1}{4}\log_2\frac{1}{4} + \frac{3}{4}\log_2\frac{3}{4}\right] = 0.81 \text{ bit}$$

Thus, for every probability distribution, we find a specific entropy (see the graph in Fig. 6.6A). We find that the maximum entropy for tossing a coin is when probability is equal ($\frac{1}{2}$) for heads and tails, which occurs when the probability distribution is uniform (scenario III, Fig. 6.6A).

Without further proof, we state that the above result may be generalized to any probability distribution: variables show maximum entropy when they are uniformly distributed. Thus, in the example in Fig. 6.5, the variable in the left panel with a uniform distribution has higher entropy than the variable with a Gaussian distribution.

In panels E and F in Fig. 6.5, we consider a 2D distribution where the observations are represented by dots in a plane. In this example we have a joint distribution similar to the one for variables a and b shown in Fig. 6.4 (where each random variable has five possible outcomes). To compute the entropy associated with such a joint probability distribution, we follow the same approach as for the 1D case: we summate $-p(x)\log_2 p(x)$ over the domain of x, which leads us to actually have two sums:

$$S(a,b) = -\sum_{i=1}^{5}\sum_{j=1}^{5} p_{a_i b_j}\log_2 p_{a_i b_j}$$

The entropy $S(a,b)$ of the joint distribution (the table in Fig. 6.4) is:

$$-\,[0.01\log_2 0.01 + 0.03\log_2 0.03 + 0.03\log_2 0.03 + \cdots + 0.03\log_2 0.03$$
$$+\, 0.02\log_2 0.02 + 0.01\log_2 0.01] = 4.29 \text{ bits}$$

The entropy for the individual variables a and b can be obtained from the marginal distributions (i.e., the five probabilities corresponding to each outcome, given one random variable). Using the distribution for a (see marginal distribution [red] in Fig. 6.4), we find that entropy $S(a)$ is:

$$-\,[0.1\log_2 0.1 + 0.2\log_2 0.2 + 0.2\log_2 0.2 + 0.4\log_2 0.4 + 0.1\log_2 0.1] = 2.12 \text{ bits}$$

and for b (marginal distribution [black] in Fig. 6.4) we find that $S(b)$ is:

$$-\,[0.1\log_2 0.1 + 0.3\log_2 0.3 + 0.3\log_2 0.3 + 0.2\log_2 0.2 + 0.1\log_2 0.1] = 2.17 \text{ bits}$$

Now we see that $S(a) + S(b) = 2.12 + 2.17 = 4.29$, which is equal to $S(a,b)$. This is not so surprising because the probability distributions for a and b were independent such that $p_{ab}([ab]) = p_a(a)p_b(b)$. If our distribution in Fig. 6.4 had been uniform, we would have found different values for the entropies. If this were the case, the five probabilities of p_a and p_b would be $\begin{bmatrix} 0.2 & 0.2 & 0.2 & 0.2 & 0.2 \end{bmatrix}$ and the joint distribution would also be uniform, with all 25 probabilities equal to 0.04. The associated entropies would now be:

$$S(a,b) = -25 \times 0.04 \times \log_2(0.04) = 4.64 \text{ bits}$$

and

$$S(a) \text{ and } S(b) \text{ are both} - 5 \times 0.2 \times \log_2(0.2) = 2.32 \text{ bits}$$

In all cases the entropies are higher (because of the uniform distribution), but,

$$S(a) + S(b) = S(a,b) \qquad\qquad (6.14a)$$

still holds.

If there were dependence between the two distributions of a and b, we would have found a different result. Let us explore the effect of dependence with an even simpler example by getting back to our coin toss. Assuming that we toss two coins 16 times, and in one case we have the usual situation where the tosses are independent (the idealized outcome is shown in Fig. 6.6B, top diagram). However, in the other case there is a "magical" full dependence between the two coins: if one coin lands on heads or tails, the other coin will too (the idealized outcome is shown in Fig. 6.6B, bottom diagram).

When tossing two coins, we have four alternative outcomes: heads/heads, heads/tails, tails/heads, and tails/tails. Assuming we have equal probability for heads and tails, we get in the independent case that the probability for each outcome is $\frac{1}{2} \times \frac{1}{2} = \frac{1}{4}$. In the fully dependent case, however, the probabilities for heads/tails and tails/heads are zero because one coin will magically copy the outcome of the other (we imagine this just for the sake of this example; do not worry about how you would actually do such a thing). The probabilities for heads/heads and tails/tails are therefore each $\frac{1}{2}$. Regardless of independence or dependence, we can first compute entropies S_1 and S_2 for each individual coin toss from the marginal distributions, finding that S_1 and S_2 are both:

$$-2 \times 0.5 \times \log_2(0.5) = 1 \text{ bit}$$

As was the case for $S(X)$ computed earlier in this chapter, it makes sense that the entropies S_1 and S_2 should be equal to 1 bit, since the outcome of each individual coin toss can fit in 1 bit (0 for heads, 1 for tails).

In the independent case (top diagram in Fig. 6.6B), we find for the joint entropy, $S_{1,2}$,

$$-4 \times 0.25 \times \log_2(0.25) = 2 \text{ bits}$$

Here we see that just as in the case for the independent distribution in Fig. 6.4, the sum of the individual entropies equals the joint entropy:

$$S_1 + S_2 = S_{1,2}$$

We can also see that the result we get for $S_{1,2}$ is reasonable since the possible joint outcomes fill four possible states, or 2 bits.

Now we compute the joint entropies for the two other scenarios in Fig. 6.6B. In the fully dependent case (the bottom diagram in Fig. 6.6B), the joint entropy $S_{1,2}$ is:

$$-\left[2 \times 0.5 \times \log_2(0.5) + 2 \times 0 \times \log_2(0)\right] = 1 \text{ bit}$$

It should be unsurprising that the joint entropy ($S_{1,2}$) in this case is identical to the entropy of an individual coin toss (S_1 or S_2); since there is total dependence, no additional information is provided by the flipping of a second coin.

The (idealized) outcomes for a case with partial dependence between the two coins are shown in the middle diagram in Fig. 6.6B. Note that due to the partial dependence, in most but not all cases, the outcomes of the first and second coin toss are identical. This results in a joint entropy of:

$$-\left[2 \times 0.125 \times \log_2(0.125) + 2 \times 0.375 \times \log_2(0.375)\right] = 1.8 \text{ bits}$$

In both cases where there is (full or partial) dependence between the tosses of the coins, we find that $S_1 + S_2 > S_{1,2}$, and the more dependence that exists between the tosses, the larger the difference between $S_1 + S_2$ and $S_{1,2}$. Accordingly, we need to adapt Equation (6.14a) to account for the possibility of dependence between the two variables by including a term that reflects this dependence. This term is commonly indicated by mutual information (MI), which is an indication of the level of dependence between variables. In other words, MI quantifies the amount of information that variable 1 provides about variable 2. In case of the dependence between coin tosses, the outcome of one coin toss determines the outcome of the other. In the normal, fair tosses the outcome of one toss does not provide any information about the other since they are independent.

To summarize, we find:

	Independent (bit)	Slightly dependent (bit)	Fully dependent (bit)
Entropy coin 1, S_1	1.0	1.0	1.0
Entropy coin 2, S_2	1.0	1.0	1.0
Sum, $S_1 + S_2$	2.0	2.0	2.0
Joint entropy, $S_{1,2}$	2.0	1.8	1.0
Difference $(S_1 + S_2) - S_{1,2}$ (=MI)	0.0	0.2	1.0

It can be seen in this example that the MI variable is indeed proportional to the level of dependence. Without further proof, we assume that we may generalize our findings and state that for any two random processes X and Y, we can compute the entropy for each of the individual processes $S(X)$ and $S(Y)$ such that their joint entropy $S(X,Y)$ is the sum of the individual entropy values when X and Y are independent (Equation (6.14a)), and otherwise:

$$S(X, Y) = S(X) + S(Y) - \text{MI}(X, Y) \tag{6.14b}$$

in which $\text{MI}(X,Y)$ is the mutual information between X and Y. Note that Equation (6.14b) holds even when X and Y are independent, since then $\text{MI}(X,Y)$ merely becomes zero. We could define joint entropy $S(X,Y)$ as the total information of the joint process X,Y. To summarize, it can be concluded from the above that **for any given pair of processes X and Y, maximal independence occurs at maximal joint entropy (or joint information) $S(X,Y)$ and minimal mutual information**

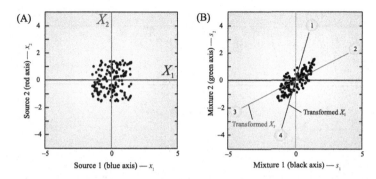

Figure 6.7 (A) Scatter plot of two source signals x_1 and x_2. (B) Scatter plot of two mixtures s_1 and s_2 that were created from the source signals. The transformed source axes (X_1, indicated by 1–4 [dark blue], and X_2, indicated by 2–3 [red]) are indicated in this mixture plot. *(Color in electronic version.)*

MI(*X,Y*) of the joint process. This is a basis for the ICA technique: independence of separated sources is evaluated by joint entropy (joint information) and MI. For the separation of independent sources, their joint information $S(X,Y)$ must be maximized (therefore, this ICA technique is also called infomax), which is the same as minimizing their mutual information MI(*X,Y*).

6.4.2   Using the Scalar Product to Find Independent Components

After we obtain the criteria for unmixing a mixture of signals (e.g., decorrelation, statistical independence, maximizing joint entropy), the procedure for separating components from mixtures in ICA is essentially the same as was outlined for PCA in Section 6.2: source signals are found from the product of the unmixing matrix and the recorded signals (Fig. 6.2). The unmixing matrix contains the vectors along which the components are extracted. **The difference between ICA and PCA is the strategy for finding the directions of the vectors in the unmixing matrix. In PCA we found directions of maximal variance (Fig. 6.3C) while the components were decorrelated. For ICA we demand statistical independence.**

To illustrate an ICA-type extraction procedure, let us consider a 2D case: two sources x_1 and x_2 creating two mixtures s_1 and s_2. Scatter plot representations of the sources and the mixtures are shown in Fig. 6.7; the sources are plotted in panel A and the resulting mixtures in panel B. Assuming that we know the mixing matrix A in this example,

$$A = \begin{bmatrix} 0.2 & 0.8 \\ 0.7 & 0.4 \end{bmatrix}$$

we can determine the orientation of the original axes X_1 and X_2 from the source scatter plot (depicted in Fig. 6.7A) in the mixture scatter plot (shown in Fig. 6.7B).

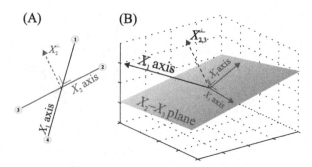

Figure 6.8 (A) and (B) show the strategy for unmixing. In (A) we have a 2D case: if we cancel all components in the direction of axis X_2 (by using the inner product with vector $X_2^\perp$ perpendicular to X_2), the remainder must be a component of the X_1 axis. This approach can be extended to higher-dimensional cases (B): by canceling components for X_2 and X_3 (by using the inner product with vector $X_{2,3}^\perp$ perpendicular to X_2 and X_3), we keep the ones for X_1. *(Color in electronic version.)*

The first axis $X_1 = [1\ 0]$, so the transformed version of source axis X_1 in the scatter plot of the mixtures is:

$$\underbrace{A}_{\text{Mixing matrix}} \underbrace{\begin{bmatrix} 1 \\ 0 \end{bmatrix}}_{X_1} = \begin{bmatrix} 0.2 & 0.8 \\ 0.7 & 0.4 \end{bmatrix} \begin{bmatrix} 1 \\ 0 \end{bmatrix} = \begin{bmatrix} 0.2 \times 1 & + & 0.8 \times 0 \\ 0.7 \times 1 & + & 0.4 \times 0 \end{bmatrix} = \begin{bmatrix} 0.2 \\ 0.7 \end{bmatrix}$$

This is the first column of mixing matrix A. Similarly, the transformed second source axis X_2 in the mixture plot is the second column of A. The axes X_1 and X_2 from the scatter plot in Fig. 6.7A are also depicted, after transformation with mixing matrix A, in Fig. 6.7B. After this transformation, X_1 becomes the axis 1−4 (dark blue) and X_2 becomes axis 2−3 (red).

For the following explanation it helps to look at the plot of the mixtures in Fig. 6.7B and the orientation of axes and vectors in Fig. 6.8A. First we establish that we know there are two sources, and that we have two mixtures. If we know the orientation of axes X_1 and X_2, we can find the contribution of x_1 to the mixtures by excluding all contributions of x_2. Because the contributions of x_2 are in the direction of axis X_2, we can use the scalar product of (1) the observation vectors of the mixture (all points in Fig. 6.7B) and (2) a vector $X_2^\perp$ perpendicular to axis X_2 (Fig. 6.8A). All components parallel to X_2, which we will indicate by $X_2^=$, will cancel since the inner product $X_2^\perp \cdot X_2^= = 0$. Therefore, the only component remaining in the scalar product of each observation in the mixture plot $[s_1\ s_2]$ with $X_2^\perp$ will be independent of x_2, and thus must be from x_1.

Note: Summarizing the above approach in a few words, **all components independent of X_2, the axis for source x_2, can only be a component of x_1.** Furthermore, we found that **we can use the inner product to remove x_2 components and only keep the ones independent from X_2.**

We can use a similar reasoning for mixtures from three or more sources. Let us consider a three-source and three-mixture case, creating a 3D space (Fig. 6.8B). If we want to find the components for x_1, we need to remove the components for x_2 and x_3. So if we construct a plane through the axes for x_2 and x_3, we can come up with $X_{2,3}^{\perp}$ (Fig. 6.8B). The inner product of observation $[s_1\ s_2\ s_3]$ with $X_{2,3}^{\perp}$ (perpendicular to the X_2–X_3 plane) removes all components associated with x_2 and x_3, and must therefore be the contribution of x_1. With a higher number of dimensions (that is, with more sources and mixtures), we can always construct a hyperplane through all-but-one selected axis (i.e., the axis of one selected source) and find a vector perpendicular to this hyperplane. This vector (analogous to $X_{2,3}^{\perp}$ in Fig. 6.8B) can then be used to remove the contributions from all directions embedded in the hyperplane (analogous to the X_2–X_3 plane in Fig. 6.8B) so that the remainder must be the contribution from the selected source.

6.4.3   A MATLAB Example

We have now set the stage for an example with two sources and two mixtures. To extract sources from the mixtures, we will follow the strategy below:

(1) We find that our mixtures are Gaussian-like distributed, but we assume that our source signals have a uniform distribution (Fig. 6.5).
(2) We use entropy to evaluate independence of the sources (Fig. 6.6).
(3) If we know the axes associated with the sources, we know how to extract a component from a mixture by using the inner product (Fig. 6.8).

Now we must deal with the fact that we (pretend that we) do not know the transformed axes for the sources in the scatter plot of the mixture (Fig. 6.7B). We will solve this problem by applying a brute force iterative approach—that is, we systematically evaluate a series of angles for source axes X_1 and X_2 and for each pair of angles we use the inner product to compute the associated sources x_1 and x_2. For every solution of x_1 and x_2 (i.e., for each direction associated with axes X_1 and X_2), we determine the level of independence of x_1 and x_2 (while we assume each satisfies a uniform distribution). We can do this in multiple ways, but for now we will evaluate how independent x_1 and x_2 are by looking at the level of mutual information $MI(x_1, x_2)$ between x_1 and x_2 (Equation (6.14b)). The more independent x_1 and x_2 are estimated to be, the lower their MI. So by following this brute force procedure, we get a series of angles for the axes X_1 and X_2 each with an associated value for $MI(x_1, x_2)$. Finally we complete our procedure by selecting the angles for axes X_1 and X_2 that correspond to the minimal value of $MI(x_1, x_2)$.

The brute force iterative procedure is followed in MATLAB script Pr6_4 *(available on http://www.elsevierdirect.com/companions/9780123849151). The following is a snippet from this script showing the iteration loops. Each iteration loop goes through a range of angles: $0-2\pi$ rad. For each loop (i.e., for each angle in the brute force search), the entropy (* H *) and mutual information (* MI *) are determined using function* entropy_2D.m *(which must be in the same directory). Due to the large number of loops in the brute force search, running this script may take ~ 30 min.*

```
%----------------------------------------------------------------
% rotate the unmixing vector and determine the mutual information of result
%----------------------------------------------------------------
MI_min=100000000000000;            % set minimum of the
                                   % mutual-information to
                                   % large number
phi_min1=0;phi_min2=0;             % set the angles for axes X1 and X2
                                   % to zero
ct_phi1=0;                         % initialize counter 1
for phi1=0:2*pi/precision:2*pi;    % LOOP for rotating axis X1
  ct_phi1=ct_phi1+1;               % update counter 1
  ct_phi2=0;                       % initialize counter 2
  for phi2=0:2*pi/precision:2*pi;  % LOOP for rotating axis X2
    ct_phi2=ct_phi2+1;             % update counter 2
    v1=[cos(phi1) sin(phi1)];      % unit vector along X1 with angle
                                   % phi1
    ic1=v1*S;                      % unmix mixture S
    ic1=ic1-mean(ic1);sigma=std(ic1); % demean and determine standard
                                   % deviation
    v2=[cos(phi2) sin(phi2)];      % unit vector along X2 with angle
                                   % phi2
    ic2=v2*S;                      % unmix mixture S
    ic2=ic2-mean(ic2);sigma=std(ic2); % demean and determine standard
                                   % deviation

% Use 2D entropy estimate function entropy_2D to compute mutual
% information (MI) and entropy (H) as a function of the
% position of axes X1 (counter for phi1) and X2 (counter for phi2)
[H(ct_phi1,ct_phi2), MI(ct_phi1,ct_phi2)]=entropy_2D(ic1,ic2);

if MI(ct_phi1,ct_phi2) < MI_min;   % TEST: current MI < current
                                   % minimum of MI ?
  MI_min=MI(ct_phi1,ct_phi2);      % if so a new minimum (minimum
                                   % for MI) is found
  imin=ct_phi1;jmin=ct_phi2;       % the indices for the new minimum
                                   % are saved
  phi_min1=phi1;                   % and so are the other relevant data
  v_min1=v1;                       % the angles, the vectors &
                                   % components
  ic1_min=ic1;
  phi_min2=phi2;
  v_min2=v2;
  ic2_min=ic2;
```

```
        end;
    end;
end;
```

Running the script Pr6_4.m (available on http://www.elsevierdirect.com/companions/ 9780123849151; also available in a low-resolution version if you are in a hurry) will show you the source and mixed signals plus their scatter plots. The scatter plot of the mixture generated by the script resembles the detail in Fig. 6.9B. The rhomboid shape of the scattered dots indicates that we are dealing with a mix of uniformly distributed source signals. For a range of combinations of hypothetical directions for axes X_1 (indicated by 1−4 in Fig. 6.9) and X_2 (indicated by 2−3 in Fig. 6.9), the MI is computed.

Note: This is a rather computationally intensive procedure associated with our brute force approach, because we (pretend that) we do not know the directions for X_1 and X_2 and just compute the MI for all directions with a precision of $1°$ for each axis, resulting in 360^2 combinations. In the low-resolution version of script Pr6_4 we compute MI every $10°$ for each axis (36^2 combinations), which greatly reduces the run time for the program.

For illustration purposes in Fig. 6.9, we kept the angle for MI associated with axis X_1 at its minimum and plotted the MI associated with axis X_2 for each angle (from $0°$ to $360°$, incremented in steps of $1°$) as a single dot and connected the dots with a line. The minima of the MI and the line connecting them are indicated in the detailed plot in Fig. 6.9B. It is obvious that the line between these minima (the double arrow indicated by $X_2^{\perp}$ in Fig. 6.9B) is a very good estimate of the vector perpendicular to axis X_2 (indicated by 2−3). The script Pr6_4 also has the option of computing the principal components; the first principal component is also indicated in Fig. 6.9B (indicated as "PCA: axis-1"; light-blue line). It is obvious that this line is in the direction of maximal variance, but it would not do a good job separating the sources in this example.

6.4.4 What If Sources Are Not Uniformly Distributed?

For the ICA examples so far, we assumed that the sources were characterized by a uniform distribution (e.g., Fig. 6.5A, C, and E) and we used entropy estimates to determine the level of independence (e.g., Fig. 6.6) of the separated candidate sources. So what should we do if we know that our sources are not uniformly distributed—take for example a human voice or a chirp—in a recording of a sound

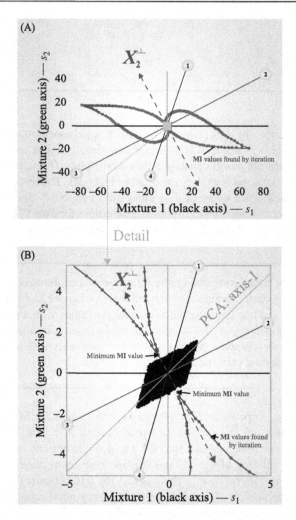

Figure 6.9 An example of an application of the ICA unmixing procedure; (B) is a detail of (A). This example shows a mixture where the sources are uniformly distributed. By iteration, we determine the MI of the candidate sources for each unmixing angle and plot this result (dots [red] in A and B). The best choice of angle is obtained when MI is minimal ("Minimum MI value" in B). Since we know, for this example, the orientation of source axis X_2 (the axis labeled at its ends by 2−3, red), we can see that our estimate for $X_2^{\perp}$ (dashed, double arrow, red) running through the minima we found for MI is indeed perpendicular to X_2. This estimate for $X_2^{\perp}$ will therefore perform a good unmixing operation. The axis labeled "PCA: axis-1" (light blue) in (B) indicates the direction of the eigenvector associated with the largest eigenvalue (the first principal component). As you can see, the principal component is not perpendicular to any of our source axes (1−4 or 2−3) and consequently would not achieve a good unmixing result for our two mixtures. The graph in (B) can be obtained with MATLAB script Pr6_4.m (available on http://www.elsevierdirect.com/companions/ 9780123849151). *(Color in electronic version.)*

mixture? Such sources usually show a distribution with many values around zero (Fig. 6.10). In such a case one could look for another function to maximize or minimize (instead of using entropy or MI). In this example of a peaky distribution (Fig. 6.10B), one could maximize for peakyness of the distribution (kurtosis, a measure for how peaky a pdf is, might do the job in such a case). Alternatively, one could transform the non-uniform distribution into a uniform one and subsequently apply the same procedures that are available for separating sources with a uniform distribution. This approach is depicted in Fig. 6.11. The example in Fig. 6.11 shows the transformation of a Gaussian distribution into a uniform one. The function used for the transformation is the cumulative probability density function (cdf). For a normally distributed variable x with zero mean, the cdf is $\frac{1}{2}\left[1 + \text{erf}(x/\sigma\sqrt{2})\right]$, in which erf is the error function (available in MATLAB) and σ is the standard deviation of x. This shows that it is plausible that for any pdf, the cdf is the optimal transformation to obtain a uniform distribution. The cdf will have the steepest slope where the probability is highest (and where you will therefore collect most observations), and this steeper slope will distribute the more densely packed observations over a wider area (see the area in between the [red] dotted lines in Fig. 6.11A−C). In contrast, at low probabilities (where fewer observations occur), the slope of the cdf will be less steep, and consequently the observations will be distributed over a smaller area. The overall effect of the transformation is thus to spread out observations more uniformly, exactly what we want for our purpose. After we transform our unmixed result into a uniform distribution, we can apply exactly the same procedure for separating mixtures (from uniformly distributed sources) we followed earlier.

The approach of transforming the source data is demonstrated in Pr6_5.m *(available on http://www.elsevierdirect.com/companions/9780123849151). The following part, with the iteration loops (similar to the ones shown above for script* Pr6_4 *), shows the transformation from the Gaussian distributed signals* ic1 *and* ic2 *to the uniformly distributed* T_ic1 *and* T_ic2.

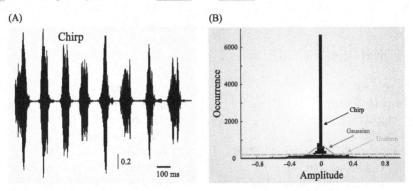

Figure 6.10 (A) Plot of a chirp (available in MATLAB by load chirp). (B) The amplitude distribution of the chirp signal compared to Gaussian (red) and uniform (blue) distributions. Note that the histogram of the chirp signal shows a peaky distribution, which is typical for many audio signals. *(Color in electronic version.)*

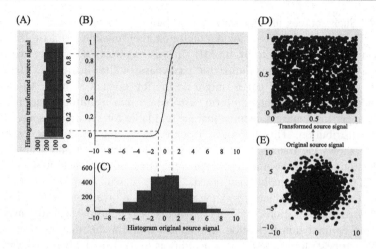

Figure 6.11 A uniform distribution (A) can be obtained from a non-uniform distribution by a transformation. This example shows a transformation of a histogram of observations drawn from a Gaussian distribution (C) using the cumulative probability density of the Gaussian distribution shown in (B). It can be seen that the majority of observations of the Gaussian distribution are located around zero in between the (red) dotted lines. Following these dotted lines from (C) to (A), it can be seen that the transformation of the Gaussian data (C) with the function in (B) distributes these points more evenly (A). Accordingly, if such a transformation is applied to a 2D scatter plot of a Gaussian variable (E), we get a scatter plot of uniformly distributed points (D). *(Color in electronic version.)*

```
% ─────────────────────────────────────────
% rotate the unmixing vector and determine the mutual information of result
% ─────────────────────────────────────────
MI_min=100000000000000;            % set minimum of the
                                   % mutual-information
                                   % to large number
phi_min1=0;phi_min2=0;             % set the angles for X1 and X2 to
                                   % zero
ct_phi1=0;                         % initialize counter 1
for phi1=0:2*pi/precision:2*pi;    % LOOP for X1
  ct_phi1=ct_phi1+1;               % update counter 1
  ct_phi2=0;                       % initialize counter 2
  for phi2=0:2*pi/precision:2*pi;  % LOOP for X2
    ct_phi2=ct_phi2+1;             % update counter 2
    v1=[cos(phi1) sin(phi1)];      % unit vector along X1 with
                                   % angle phi1
    ic1=v1*S;                      % unmix mixture S
```

```
        ic1=ic1-mean(ic1);sigma1=std(ic1);  %  demean and determine standard
                                            %  deviation
        v2=[cos(phi2) sin(phi2)];           %  unit vector along X2 with angle
                                            %  phi2
        ic2=v2*S;                           %  unmix mixture S
        ic2=ic2-mean(ic2);sigma2=std(ic2);  %  demean and determine standard
                                            %  deviation
        % TRANSFORMATION to make the estimates uniformly distributed
        % Transform ic1 and ic2 to a uniform distribution using erf to
        % transform the demeaned ic1 and ic2
        T_ic1=0.5*(1+erf((ic1)/(sigma1*sqrt(2))));
        T_ic2=0.5*(1+erf((ic2)/(sigma2*sqrt(2))));

        % Use custom estimate function entropy_2D to compute mutual
        % information (MI) and entropy (H)
        [H(ct_phi1,ct_phi2), MI(ct_phi1,ct_phi2)]=entropy_2D(T_ic1,T_ic2);

        % Sum of the variances should max at independence
        sum_var(ct_phi1,ct_phi2)=std(ic1)^2+std(ic2)^2;

        if MI(ct_phi1,ct_phi2) < MI_min;    %  TEST: current MI < current
                                            %  minimum of MI ?
          MI_min=MI(ct_phi1,ct_phi2);       %  if so a new minimum (maximum
                                            %  for MI) is found
          imin=ct_phi1;jmin=ct_phi2;        %  the indices for the new minimum
                                            %  are saved
          phi_min1=phi1;                    %  and so are the other relevant data
          v_min1=v1;                        %  the angles, the vectors &
                                            %  components
          ic1_min=ic1;
          phi_min2=phi2;
          v_min2=v2;
          ic2_min=ic2;
        end;
      end;
    end;
```

Because script Pr6_5.m is also computationally demanding, there is a low-resolution version available as well (both available on http://www.elsevierdirect.com/companions/9780123849151). Running the script will show you the source signals and the mixtures with their associated scatter plots. The scatter plot of the mixtures will resemble Fig. 6.12 and will also show the transformed axes X_1 (dark blue, indicated by 1−4 in Fig. 6.12) and X_2 (red, indicated by 2−3 in Fig. 6.12) plus the estimated MI

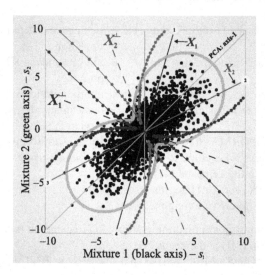

Figure 6.12 The example in this figure is the same as in Fig. 6.9, except now the source signals are normally distributed. As in the previous examples in Figs. 6.7–6.9, the two source axes are X_1 (labeled at its ends by 1–4, dark blue) and X_2 (labeled 2–3, red). The dashed, double arrows $X_1^{\perp}$ (dark blue) and $X_2^{\perp}$ (red) indicate the directions for minimal MI found by iteration. The MI values we computed by iteration are indicated by the dots interconnected by lines: dark blue for $X_1^{\perp}$ and red for $X_2^{\perp}$. The axis labeled "PCA: axis-1" (light blue) is the direction of the eigenvector associated with the largest eigenvalue (the first principal component). The 8-shaped contour (light-blue dots) denotes the variance for each direction. Further details can be found in the text. *(Color in electronic version.)*

values associated with each axis (red and dark-blue dots interconnected with lines in Fig. 6.12). The program Pr6_5.m also allows us to compute the principal components (the first component, "PCA: axis-1," is indicated in Fig. 6.12) and the variance associated with each angle (the 8-shaped contour, light blue in Fig. 6.12). Note again how the first PCA component is not orthogonal to any of the source axes X_1 and X_2.

The implicit underlying thought of the algorithm in Pr6_5.m is that by transforming the unmixed sources into a uniform distribution, we can follow the same procedure as earlier without affecting the information content. This assumption may seem like a bit of a stretch, but if we transform the data using an invertible function (such as our cumulative probability density function in Fig. 6.11B), we do not affect the mutual independence of the signals (see Stone, 2004).

An invertible function is defined as a function that creates a unique new data point for each original data point and (because the function is invertible) this transformation can also be reversed. This means that if we have several independent data sets, they will remain independent after transformation with the invertible function into the other domain and vice versa.

6.4.5   Can We Apply Smarter Approaches Than the Brute Force Technique?

In the above brute force approach we looked into a 2D case. We determined the source axes X_1 and X_2 with a precision of $1°$; this corresponds to 360 computations for each axis. For the 2D case this evaluates to $360^2 = 129{,}600$ iterations. In other words, for each iteration, we guess candidate sources and we compute their MI. For more dimensions and/or higher precisions, the number of iterations grows rapidly—for example, if we wanted a $\frac{1}{2}°$ precision in a six-source case we have $720^6 \approx 1.4 \times 10^{17}$ iterations. As you can see, we need a more efficient procedure to find the best angles for the source axes; otherwise, source extraction very rapidly becomes a computational nightmare. Let us look at the surface of the inverse of the MI. This is the approach we take in MATLAB script Pr6_6.m, which is almost identical to Pr6_4.m, but now we use the inverse of the MI instead of the MI itself for visualization reasons; minima in MI are maxima in 1/MI, and maxima are easier to show in a 3D plot (both scripts are available on http://www.elsevierdirect.com/companions/9780123849151). Such a plot generated by Pr6_6.m for our two-mixtures/two-sources scenario is depicted in Fig. 6.13. The horizontal axes are the angles for axes X_1 and X_2, and the vertical axis is 1/MI. There are clearly eight maxima in the 1/MI landscape (again, corresponding to minima of MI) present in the plot. These eight maxima are not surprising if we consider the example of the source axes in the mixture scatter plot in Fig. 6.7B. Each source axis is labeled at each end (1 and 4 for axis X_1 and 2 and 3 for axis X_2), since each axis can be characterized twice: by its optimal angle ϕ or by the same angle plus

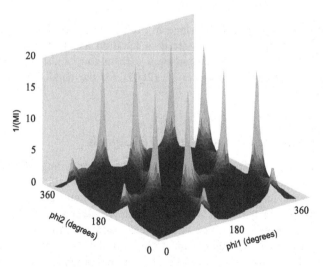

Figure 6.13 The inverse of the MI (1/MI) of two components extracted from two mixtures as a function of the angles (phi1, phi2) of the two source axes. Minima for the MI show up as maxima in 1/MI surface. The eight maxima correspond to a pair of angles of the axes that extract sources with minimum MI from the mixtures. To find these two axes, we need to identify only one of the eight maxima. Therefore, we can use the gradient in the landscape to locate one of the maxima. This procedure is more efficient than a brute force computation of the whole 1/MI surface. This graph was obtained with Pr6_6.m (available on http://www. elsevierdirect.com/companions/9780123849151). *(Color in electronic version.)*

$180°$ ($\phi + 180°$). An optimal combination (associated with a maximum $1/MI$) is any combination of the angles for which we can successfully unmix our mixture into its source components. Using the labels in Fig. 6.9, we find that we have eight of these combinations: 1,2 1,3 2,1 2,4 3,1 3,4 4,2 4,3. Note that the combinations 1,4 and 2,3 are not valid because they denote the same axis and not a pair of axes. The eight combinations correspond to the eight maxima in the $1/MI$ landscape in Fig. 6.13. Because each of the eight combinations describes the optimal angles for the two axes, we only need to find one pair (i.e., one maximum in the landscape in Fig. 6.13) for our unmixing procedure. Since we find that the landscape of $1/MI$ has a clear-cut structure, we can use this to our benefit and avoid a lengthy brute force computation. Instead of computing the values for all angles from $0-360°$, we apply the so-called simplex method, which uses the gradient in the landscape to locate one of the maxima. Globally, this procedure works as follows:

(1) We randomly pick an initial point in the $1/MI$ landscape (a pair of angles ϕ_1 and ϕ_2 associated with our pair of source axes X_1 and X_2) and determine its $1/MI$ value.
(2) Then we pick two neighboring points in the landscape, and determine $1/MI$ for these points as well (now we have defined a triangle in the landscape).
(3) We determine which of the three points has the lowest $1/MI$ value and move it in the direction where $1/MI$ is largest.
(4) We repeat Step 3 above until we cannot find a neighboring point with a low value of $1/MI$, at which point we conclude that we have reached a peak in the landscape.
(5) Finally we will make our triangle of points smaller and repeat Steps 3 and 4 to locate the maximum with optimal precision.

By using this method, we use the slope in the landscape to climb toward a maximum, a procedure that is much faster than iteration and that scales much better when we increase the number of sources in each mixture and the number of measurements of mixtures. See Press et al. (2007) for the details of this and other parameter search techniques.

6.4.6   An Example of ICA Applied to EEG Signals

The signals of brain electrical activity in Fig. 6.1 are recordings directly from the cortex (ECoG) and from the scalp (EEG). In the context of this chapter, it is fairly reasonable to assume that the signals generated at different locations separated by several millimeters in the brain will be statistically independent. Another way of saying the same thing is that since brain signals carry a lot of information, recording sites that are relatively remote must have low levels of MI. When recording directly from the cortex, we can indeed observe this principle. When we use ICA to decompose an ECoG (Fig. 6.1A), our statistically independent components are almost identical to the recorded channels, indicating that the different sites on the cortex generate statistically independent signals. For the EEG (Fig. 6.1B), this reasoning does not hold because the skull and scalp have a tendency to smear (mix) the contributions of the underlying sources, so ICA may be a good tool to find individual sources in the signals.

The procedure that is commonly applied to EEG analysis is to find source signals that are temporally independent (because the EEG matrix has a temporal and spatial component, we could also look for components that are spatially independent, but this is usually not done). The underlying thought here is that source signals contribute to the signal at each EEG electrode. An example of two sources contributing to

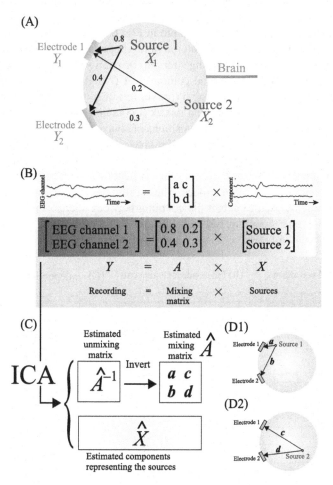

Figure 6.14 ICA analysis in EEG, a two-source and two-channel example. (A) Electrodes Y_1 and Y_2 record channels 1 and 2, each containing a mixture of sources X_1 and X_2. In each mixture, the attenuation of the source signal is proportional with distance between source and electrode (symbolized by the arrows). (B) shows the mathematics underlying the mixing process that can be represented by matrix multiplication $Y = AX$, with A being the mixing matrix (similar to the example in Fig. 6.2). The next step, (C), is to estimate the mixing matrix and source components with the ICA procedure. The estimated source activity can give an impression of the distribution of activities across the brain and the estimate of the mixing matrix can be used to determine the effect for each source on the electrodes (D). *(Color in electronic version.)*

two EEG electrodes/channels is depicted in Fig. 6.14A. The EEG data can therefore be considered a linear mixture of the sources. Because the electrodes register the fields of the locally generated activity traveling at the speed of light, the delays for propagation between source and electrode are negligible. Our finding with the ECoG (that the ICA components resemble the original time series) shows that if sources are not too close, they will be independent. The simplified scenario in Fig. 6.14A shows how two electrodes Y_1 and Y_2 each record a different mixture from sources X_1 and X_2. Similar to the example in Fig. 6.2, we have that $Y = AX$, with A being the mixing matrix. The first column in mixing matrix A (a and b, in the example 0.8 and 0.4) indicate the coupling strength (proximity) of source X_1 to electrodes Y_1 and Y_2 (Fig. 6.14B). The second column in A (c and d, in the example 0.2 and 0.3) reflect the same coupling strength (proximity) of source X_2 to electrodes Y_1 and Y_2 (Fig. 6.14B). In this sense, mixing matrix A contains spatial information because the values of its elements reflect the positions of sources and electrodes.

Now we can use ICA to estimate our source signals $\hat{X}$ and the unmixing matrix $\hat{A}^{-1}$ (Fig. 6.14C). Assuming that $\hat{A}^{-1}$ is invertible, we can determine an estimate $\hat{A}$ of the mixing matrix. Matrix $\hat{A}$ contains the estimates for coupling strengths between each of the sources and the electrodes. These estimates a, b, c, and d in Fig. 6.14C and D can now be used to depict the coupling between sources and electrodes. For Source 1 we find coupling strengths a and b for Electrode 1 and Electrode 2 (Fig. 6.14D1); for Source 2 we have strengths c and d for Electrode 1 and

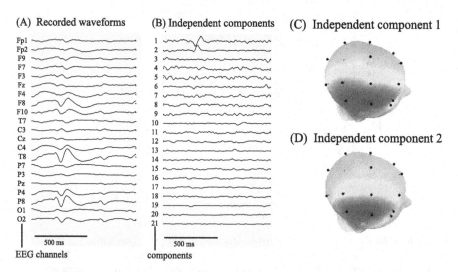

Figure 6.15 Part of the EEG recording shown in Fig. 6.1B is shown in (A); the 21 independent components are shown in (B). Here it can be seen that the epileptic spike waveforms are only represented in the first two independent components. Topographic maps of the scalp potential associated with these two components are shown in (C) and (D). These distributions are indicative for a source that is located right temporally. This figure was prepared with eeglab software. This MATLAB-based package can be downloaded from the Web site http://sccn.ucsd.edu/~scott/ica.html. *(Color in electronic version.)*

Electrode 2 (Fig. 6.14D2). As we will demonstrate in the following example (because usually the EEG recording includes multiple channels), it is common practice to show the coupling strength for each component (source) at each electrode in a color-coded fashion.

A detail of the EEG recording in Fig. 6.1 (the epoch in between the asterisks in Fig. 6.1B) is shown in Fig. 6.15A. This EEG recording contains a high-amplitude epileptic spike. The ICA of this 21-channel recording shows two components that seem associated with this spike signal (Components 1 and 2 in Fig. 6.15B). Since each electrode corresponds to a location, we can use the multichannel EEG to map our independent components topographically (as we did in Fig. 6.14D). The topographic maps of both these independent sources show a right temporal location (Fig. 6.15C and D). It was confirmed clinically that the epileptic focus was indeed located in the right temporal lobe in this patient. Although this confirmation is reassuring, it should be noted here that the brain area where epileptic spikes are generated and the focus where the epileptic seizures originate are not always the same location.

Appendix 6.1

Eigenvalues and Eigenvectors

The eigenvalues and eigenvectors of a matrix play a role in multiple applications, including the determination of principal components described in this chapter. "Eigen" is a German word that, in this context, may be translated into "characteristic." Recall that a matrix can be used to efficiently represent a set of expressions—for example,

$x + 6y$

$5x + 2y$

can be written as the product of a matrix A and vector v:

$$\underbrace{\begin{bmatrix} 1 & 6 \\ 5 & 2 \end{bmatrix}}_{A} \underbrace{\begin{bmatrix} x \\ y \end{bmatrix}}_{v} = Av$$

Vector v is an eigenvector of A if multiplication with matrix A scales it by a constant λ without changing the direction of v (Fig. A6.1):

$$Av = \lambda v \qquad\qquad\qquad (A6.1.1)$$

The constant λ is the so-called eigenvalue of A. We can rewrite this expression as:

$$Av - \lambda v = 0 \rightarrow$$

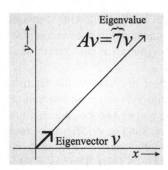

Figure A6.1 Eigenvector $v = (1,1)$ of matrix A. Note that the product Av does not change the direction of v; it only scales it by the eigenvalue λ (7 in this example). This is essentially the property associated with eigenvectors and eigenvalues of a matrix A, here v and λ, respectively. *(Color in electronic version.)*

$$(A - \lambda I)v = 0$$

where I is the identity matrix.

This expression always has the trivial solution $v = 0$, while according to Cramer's rule (see, e.g., Jordan and Smith, 1997), non-trivial solutions (solutions for which $v \neq 0$) can only exist if:

$$|A - \lambda I| = 0 \qquad\qquad (A6.1.2)$$

The $|\ldots|$ indicates the determinant of the matrix. If we go back to our numerical example above, and apply the condition in Equation (A6.1.2), we have:

$$\left| \underbrace{\begin{bmatrix} 1 & 6 \\ 5 & 2 \end{bmatrix}}_{A} - \lambda \underbrace{\begin{bmatrix} 1 & 0 \\ 0 & 1 \end{bmatrix}}_{I} \right| = 0 \rightarrow$$

$$\left| \underbrace{\begin{bmatrix} 1-\lambda & 6 \\ 5 & 2-\lambda \end{bmatrix}}_{A - \lambda I} \right| = 0 \qquad\qquad (A6.1.3)$$

This leads to the characteristic equation:

$$(1 - \lambda) \times (2 - \lambda) - 5 \times 6 = 0 \rightarrow \lambda^2 - 3\lambda - 28 = 0 \rightarrow (\lambda - 7)(\lambda + 4) = 0$$

Thus, for our numerical example we find $\lambda_1 = 7$ and $\lambda_2 = -4$.

If we generalize the matrix, let us say,

$$A = \begin{bmatrix} a & b \\ c & d \end{bmatrix}$$

the characteristic equation becomes:

$$(a - \lambda)(d - \lambda) - bc = 0 \rightarrow \lambda^2 - (a + d)\lambda + (ad - bc) = 0$$

The solutions for the eigenvalues are now:

$$\lambda_{1,2} = \frac{(a + d) \pm \sqrt{(a + d)^2 - 4(ad - bc)}}{2}$$

Once the eigenvalues are known, we can compute an eigenvector for each eigenvalue. The only thing we need to determine is the direction of the eigenvector, since the length is unimportant (the scalability, as shown in Fig. A6.1, holds for any length of vector as long as the direction is correct). So we can set the value of x in the eigenvector v arbitrarily to 1 and we substitute for the eigenvalue $\lambda_1 = 7$ in Equation (A6.1.1):

$$Av = \lambda v \rightarrow \underbrace{\begin{bmatrix} 1 & 6 \\ 5 & 2 \end{bmatrix}}_{A} \underbrace{\begin{bmatrix} 1 \\ y \end{bmatrix}}_{v} = \underbrace{7}_{\lambda} \underbrace{\begin{bmatrix} 1 \\ y \end{bmatrix}}_{v}$$

This results in two equations:

$$1 + 6y = 7$$
$$5 + 2y = 7y$$

Both have the same solution, $y = 1$. Therefore,

$$v = \begin{bmatrix} x \\ y \end{bmatrix} = \begin{bmatrix} 1 \\ 1 \end{bmatrix}$$

is an eigenvector for eigenvalue 7; this is the eigenvector shown in Fig. A6.1. The same approach can be followed for the other eigenvalue, -4.

7 Causality

7.1 Introduction

In the previous chapters we decomposed multichannel data into its components and we saw that this can be useful to detect structure in complex data sets. Another question that is often posed concerns the causal structure between channels or components: that is, does one channel generates another? In neuroscience the underlying question is: does an area in the brain activate other ones? A typical example is when an epileptologist examines multichannel recordings of brain electrical activity and attempts to find the source (focus) of the epileptic seizures. Often this task is accomplished by finding the signals that lead or lag; the leading signals are then considered as causing the lagging ones. Cross-correlation or nonlinear equivalents can be used to formalize and quantify timing differences between signal pairs in multichannel data sets (see van Drongelen, 2007, chapter 8 to review cross-correlation).

We have to start pessimistically by pointing out that translation from lead−lag to causality is strictly not possible—the example in Fig. 7.1 demonstrates this. If we record from areas A and B in Fig. 7.1A, our method of interpreting lead−lag as a causal relationship A→B is correct. However, if we measure signals from A, B, and C (Fig. 7.1B), we conclude that A→B, A→C, and B→C. We are only partly correct: the two former relationships are correctly inferred but the latter is not. It even would get worse if we had not recorded from A in this example: then we only find B→C and we would be 100% incorrect. So equating lead−lag with causality/connectivity can be incorrect. Having said this, in many studies in neuroscience this is (conveniently) ignored, and timing in signals is frequently used as an argument for connectivity. Often, authors use terms such as "functional connectivity" or "synaptic flow" to (implicitly) indicate the caveats above. Because we often know the typical conduction velocity and delays caused by synaptic transmission, we can (at least) recognize unrealistic delays. For example, if we know that areas B and C in Fig. 7.1B are 10 cm apart and that conduction velocities of the fibers between B and C are ~1 m/s, we can expect delays ~100 ms plus a few milliseconds for each synapse involved. Now suppose that in this example the delay between B and C ($\Delta t_2 - \Delta t_1$) is ~5 ms: such a value is far below the expected delay of over 100 ms, which is an indication that direct connectivity does not play a role in the observed lead−lag between B and C.

Signal Processing for Neuroscientists, A Companion Volume. DOI: 10.1016/B978-0-12-384915-1.00007-3

(A)

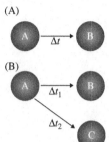

(B)

Figure 7.1 Example of areas in the brain A, B, C and their connections, indicated by the arrows. The delays, due to conduction velocity in the connections, between these areas are Δt, Δt_1, Δt_2 with $\Delta t_2 > \Delta t_1$. (A) Due to the delay, activity in area B lags relative to the activity in A. (B) Both B and C lag relative to A, but since $\Delta t_2 > \Delta t_1$, activity in C also lags relative to B.

7.2 Granger Causality

In the 1950s, Norbert Wiener proposed that one signal causes another if your pre-diction of the latter signal improves by including knowledge of the former. About a decade later, Clive Granger (1969) formalized this concept for linear regression of stochastic processes. To explain the principle, we consider an example of two sig-nals x and y, and we suppose that we can characterize them with the following autoregressive (AR) models:

$$x_n = a_1 x_{n-1} + a_2 x_{n-2} + \cdots + a_i x_{n-i} + \cdots + b_1 y_{n-1} + \cdots + b_i y_{n-i} + \cdots + ex_n$$
$$(7.1a)$$

$$y_n = c_1 y_{n-1} + c_2 y_{n-2} + \cdots + c_i y_{n-i} + \cdots + d_1 x_{n-1} + \cdots + d_i x_{n-i} + \cdots + ey_n$$
$$(7.1b)$$

Here $a_{...}$, $b_{...}$, $c_{...}$, and $d_{...}$ are the coefficients and ex_n and ey_n are error terms. If the variance of error ex_n is reduced by including any value of y, y_{n-i} (i.e., $b_i \neq 0$), then we say that y causes x. Similarly, if the variance of error ey_n is reduced by including any value of x, x_{n-i} (i.e., $d_i \neq 0$), then we say that x causes y. In this example, if $b_i \neq 0$ and/or $d_i \neq 0$, we may use the term "Granger causality" to describe the relationship between x and y; we add the qualification "Granger" to indicate that we may not be dealing with a true causal relationship. Similar to the examples given above and in Section 1.3 (where we used the Hilbert transform to detect lead and lag between channels), there may be alternative explanations why x helps to predict y or vice versa.

7.3 Directed Transfer Function

In the following we will link the Granger causality to the z-domain and frequency domain approach of the directed transfer function (DTF), first described by Kamiński and Blinowska (1991). The explanation here will be informal; a formal proof of the relationship between Granger causality and DTF is presented in Kamiński et al. (2001).

7.3.1 Autoregression in the Frequency Domain

To link time and frequency domains, we start from a 1D AR model and its frequency domain presentation. Next, we extend the approach to a multidimensional model and define the DTF. Because we want to follow the derivation by Kamiński and Blinowska (1991), we start our explanation in the time domain and transform our findings into the frequency domain via the z-transform.

7.3.1.1 1D Example

A one-channel model of order p can be represented by:

$$x_n = a_1 x_{n-1} + a_2 x_{n-2} + \cdots + a_i x_{n-i} + \cdots + a_p x_{n-p} + e_n \tag{7.2a}$$

Here x_n are the signal samples, a_i are the AR model's coefficients, and e_n is the error term, which we represent by zero mean GWN.

Equation (7.2a) can be rewritten in the form:

$$x_n - a_1 x_{n-1} - a_2 x_{n-2} - \cdots - a_i x_{n-i} - \cdots - a_p x_{n-p} = e_n \tag{7.2b}$$

Because we deal with a discrete time system, we can transform from the time domain into the z-domain (to review the z-transform, see chapter 9 in van Drongelen, 2007). We define the following transform pairs:

$$
\begin{aligned}
x_n &\leftrightarrow X(z) \\
e_n &\leftrightarrow E(z)
\end{aligned}
$$

Applying the z-transform to Equation (7.2b) we get:

$$X(z) - a_1 z^{-1} X(z) - a_2 z^{-2} X(z) - \cdots - a_i z^{-i} X(z) - \cdots - a_p z^{-p} X(z) = E(z) \tag{7.2c}$$

$$\rightarrow X(z)[1 - a_1 z^{-1} - a_2 z^{-2} - \cdots - a_i z^{-i} - \cdots - a_p z^{-p}] = E(z)$$

$$\rightarrow X(z) = \underbrace{\left[\frac{1}{1 - a_1 z^{-1} - a_2 z^{-2} - \cdots - a_i z^{-i} - \cdots - a_p z^{-p}} \right]}_{H(z)} E(z) = H(z)E(z) \tag{7.2d}$$

In Equation (7.2d), we defined $H(z)$ as the transfer function of the system with input noise e_n and output signal x_n.

Recall that the z-transform is the Laplace transform applied to discrete time series (chapter 9 in van Drongelen, 2007); the complex variable z for a time series sampled with interval Δ is defined as $z = e^{s\Delta}$, where s is the complex variable of the Laplace transform. To get from the z-domain to the frequency domain, we now use the imaginary part of exponent $s = \sigma + j\omega = \sigma + j2\pi f$ of variable z—that is:

$$z^{-1} = e^{-s\Delta} = e^{-(\sigma + j2\pi f)\Delta} \rightarrow z^{-1} = e^{-j2\pi f\Delta}$$

Using this, we can rewrite Equation (7.2d) as a function of frequency f:

$$X(f) = H(f)E(f) \tag{7.2e}$$

Now we have an expression in the frequency domain, where $X(f)$ is the Fourier transform of signal x_n, $E(f)$ is the Fourier transform of the noise input e_n, and $H(f)$ is the frequency response characterizing the system with input e_n and output x_n. Since $X(f)$ is the discrete Fourier transform of x_n, the unscaled power spectrum S for x_n (chapter 7 in van Drongelen, 2007) is:

$$S(f) = X(f)X(f)^* \tag{7.3a}$$

The * indicates the complex conjugate. Equation (7.3a) combined with (7.2e) gives:

$$S(f) = [H(f)E(f)][H(f)E(f)]^*$$

Since this expression is the product of four complex numbers (for each frequency), we may remove the brackets and rearrange the terms:

$$S(f) = H(f)E(f)E(f)^*H(f)^* \tag{7.3b}$$

If the input process e_n is zero mean GWN, its unscaled power spectrum $E(f)E(f)^* = N\sigma^2$, where N is the number of measurements of x_n, and σ^2 is the variance of the noise process.

Note: The equality $E(f)E(f)^* = N\sigma^2$ is directly related to Parseval's theorem (appendix 7.1 in van Drongelen, 2007), stating that the energy of a signal e in the time domain equals the energy of its power spectrum: $\sum(EE^*/N) = \sum e^2$. If the signal has zero mean we may use the following (biased) expression for variance: $\sigma^2 = (1/N)\sum e^2$. Combining the two latter expressions, we get a formula for the sum of the bins in the power spectrum: $\sum(EE^*/N) = N\sigma^2$. Now, if signal e is GWN, the power is equally distributed across the N bins of its spectrum—in other words, the power in each bin (the power for each frequency f) of the normalized spectrum is $N\sigma^2/N = \sigma^2$. Finally, for the non-normalized spectrum EE^* (instead of the normalized one EE^*/N) we find that the value for each frequency bin is $N \times \sigma^2 : E(f)E(f)^* = N\sigma^2$.

If we substitute this result in Equation (7.3b), we get:

$$S(f) = N\sigma^2[H(f)H(f)^*] = N\sigma^2|H(f)|^2 \tag{7.3c}$$

Thus, in this case the spectrum $S(f)$ is proportional with the power of the frequency response $|H(f)|^2 = [H(f)H(f)^*]$. Thus, in the frequency domain $H(f)$ relates input with output—that is, the noise with the signal.

7.3.1.2 Multidimensional Example

The next step is to generalize the above from a one-channel pth-order AR process to a k-channel one. As a first step let us consider a three-channel data set with a pth-order AR process. For this system, the equivalent for Equation (7.2a) becomes:

$$\underbrace{\begin{pmatrix} x(1) \\ x(2) \\ x(3) \end{pmatrix}_n}_{\vec{x}_n} = \underbrace{\begin{pmatrix} a_{11} & a_{12} & a_{13} \\ a_{21} & a_{22} & a_{23} \\ a_{31} & a_{32} & a_{33} \end{pmatrix}_1}_{A_1} \underbrace{\begin{pmatrix} x(1) \\ x(2) \\ x(3) \end{pmatrix}_{n-1}}_{\vec{x}_{n-1}} + \underbrace{\begin{pmatrix} a_{11} & a_{12} & a_{13} \\ a_{21} & a_{22} & a_{23} \\ a_{31} & a_{32} & a_{33} \end{pmatrix}_2}_{A_2} \underbrace{\begin{pmatrix} x(1) \\ x(2) \\ x(3) \end{pmatrix}_{n-2}}_{\vec{x}_{n-2}}$$

$$+ \cdots + \underbrace{\begin{pmatrix} a_{11} & a_{12} & a_{13} \\ a_{21} & a_{22} & a_{23} \\ a_{31} & a_{32} & a_{33} \end{pmatrix}_i}_{A_i} \underbrace{\begin{pmatrix} x(1) \\ x(2) \\ x(3) \end{pmatrix}_{n-i}}_{\vec{x}_{n-i}} + \cdots$$

$$+ \underbrace{\begin{pmatrix} a_{11} & a_{12} & a_{13} \\ a_{21} & a_{22} & a_{23} \\ a_{31} & a_{32} & a_{33} \end{pmatrix}_p}_{A_p} \underbrace{\begin{pmatrix} x(1) \\ x(2) \\ x(3) \end{pmatrix}_{n-p}}_{\vec{x}_{n-p}} + \underbrace{\begin{pmatrix} e(1) \\ e(2) \\ e(3) \end{pmatrix}_n}_{e_n} \tag{7.4}$$

or

$$\vec{x}_n = A_1\vec{x}_{n-1} + A_2\vec{x}_{n-2} + \cdots + A_i\vec{x}_{n-i} + \cdots + A_p\vec{x}_{n-p} + \vec{e}_n$$

Here vector:

$$\vec{x}_n = \begin{pmatrix} x(1) \\ x(2) \\ x(3) \end{pmatrix}_n$$

are the samples for the three channels $x(1)-x(3)$, the 3×3 matrix:

$$A_i = \begin{pmatrix} a_{11} & a_{12} & a_{13} \\ a_{21} & a_{22} & a_{23} \\ a_{31} & a_{32} & a_{33} \end{pmatrix}_i$$

are AR model's coefficients, and vector:

$$\vec{e}_n = \begin{pmatrix} e(1) \\ e(2) \\ e(3) \end{pmatrix}_n$$

are the errors for each channel. Note that we use arrows to indicate that x and e are vectors instead of scalars now. A useful feature is that the **AR coefficients A_i relate the x values across time and more importantly (in this context) across channels**. For instance, for:

$$A_2 = \begin{pmatrix} a_{11} & a_{12} & a_{13} \\ a_{21} & a_{22} & a_{23} \\ a_{31} & a_{32} & a_{33} \end{pmatrix}_2$$

we have $(a_{11})_2 \times x(1)_{n-2}$ relating past values of $x(1)$ (i.e., the value of $x(1)$ at sample $n-2$) to the present value of $x(1)$ (i.e., its value at sample n); $(a_{12})_2 \times x(2)_{n-2}$ relating past values of $x(2)$ (i.e., the value at sample $n-2$) to the present value of $x(1)$ (i.e., the value at n); $(a_{13})_2 \times x(3)_{n-2}$ relating past values of $x(3)$ (i.e., the value at $n-2$) to the present value of $x(1)$ (i.e., the value at n); $(a_{21})_2 \times x(1)_{n-2}$ relating past values of $x(1)$ (i.e., the value at $n-2$) to the present value of $x(2)$ (i.e., the value at n), etc. Here we can see the relationship between this approach and Granger causality: if one of the coefficients $a_{ij} \neq 0$ (with $i \neq j$), there is a causal relationship (following the definition of Granger causality as discussed in Section 7.2) between the channels i and j, such that $j \rightarrow i$. In the numerical example above, if $(a_{12})_2 \neq 0$, we find that $x(2) \rightarrow x(1)$ (i.e., a causal relationship between channels $2 \rightarrow 1$).

Following our approach for Equation (7.2a), Equation (7.4) can be rewritten in the form:

$$\vec{x}_n - A_1\vec{x}_{n-1} - A_2\vec{x}_{n-2} - \cdots - A_i\vec{x}_{n-i} - \cdots - A_p\vec{x}_{n-p} = \vec{e}_n \tag{7.5a}$$

We now repeat the same procedure as we employed for the one-dimensional case, and get to the frequency domain via the z-transform. First we define the following transform pairs:

$$\vec{x}_n \ \leftrightarrow \ \vec{X}(z)$$
$$\vec{e}_n \ \leftrightarrow \ \vec{E}(z)$$

The z-transform of Equation (7.5a) is:

$$\vec{X}(z) - A_1z^{-1}\vec{X}(z) - A_2z^{-2}\vec{X}(z) - \cdots - A_iz^{-i}\vec{X}(z) - \cdots - A_pz^{-p}\vec{X}(z) = \vec{E}(z) \tag{7.5b}$$

$$\to \vec{X}(z)[I - A_1 z^{-1} - A_2 z^{-2} - \cdots - A_i z^{-i} - \cdots - A_p z^{-p}] = \vec{E}(z)$$

$$\to \vec{X}(z) = \underbrace{\left[\frac{I}{I - A_1 z^{-1} - A_2 z^{-2} - \cdots - A_i z^{-i} - \cdots - A_p z^{-p}} \right]}_{H(z)} \vec{E}(z) = H(z)\vec{E}(z) \quad (7.5c)$$

In the above, I is the identity matrix and $H(z)$ is defined as the transfer function matrix between the input noise:

$$\vec{E}(z) = \begin{pmatrix} E(1) \\ E(2) \\ E(3) \end{pmatrix}_z$$

and output signal:

$$\vec{X}(z) = \begin{pmatrix} X(1) \\ X(2) \\ X(3) \end{pmatrix}_z$$

Similar to the procedure we followed in the 1D example above, we now use the imaginary part of exponent $s = \sigma + j\omega = \sigma + j2\pi f$ of variable z to get to the frequency domain—that is:

$$z^{-1} = e^{-s\Delta} = e^{-(\sigma + j2\pi f)\Delta} \to z^{-1} = e^{-j2\pi f\Delta}$$

We then rewrite Equation (7.5c) as a function of frequency f:

$$\vec{X}(f) = H(f)\vec{E}(f) \tag{7.5d}$$

The unscaled power spectrum S for $\vec{x}_n$ is:

$$S(f) = \vec{X}(f)\vec{X}(f)^* \tag{7.6a}$$

The * indicates the adjoint, **both** the complex conjugate and the transpose. Note that for each frequency f in Equation (7.6a) S is a 3×3 matrix. Equation (7.6a) combined with Equation (7.5d) gives:

$$S(f) = [H(f)\vec{E}(f)][H(f)\vec{E}(f)]^* = H(f)\vec{E}(f)\vec{E}(f)^* H(f)^* \tag{7.6b}$$

In the above, we used the identity $[H(f)\vec{E}(f)]^* = \vec{E}(f)^* H(f)^*$. If the input noise processes $\vec{e}_n$ are independent white Gaussian with zero mean and variance σ^2, we get

$\vec{E}(f)\vec{E}(f)^* = N\sigma^2 I$, where N is the number of measurements of $\vec{x}_n$ and I is the identity matrix. This generates:

$$S(f) = N\sigma^2[H(f)H(f)^*]$$ (7.6c)

Thus, in this case the spectrum $S(f)$ is proportional with the power of the frequency response $[H(f)H(f)^*] = |H(f)|^2$. Thus, in the frequency domain $H(f)$ relates input with output, and it is by definition inversely proportional to $A(f)$ (see Equation (7.5c)).

7.3.1.3 The Directed Transfer Function

Unlike in Equation (7.3c), H in the multichannel version in Equation (7.6c) is not a single value but a matrix for each frequency f (in this example a 3×3 matrix). **Each element H_{ij} in H represents a transfer value between channels j and i.** Again we can see the relationship between this approach and Granger causality: if the transfer value between j and i is nonzero, there is an input–output (causal) relationship. Kamiński and Blinowska (1991) use a normalized version of H, which they call the DTF, to study interrelationships between channels in their data sets. They normalize each component H_{ij} by dividing it by the sum of all elements of H in the same row of the H matrix. Because H_{im} represents the effect of channel m on channel i, you can, in a K-channel recording, normalize by division by the contributions from all channels: $\sum_{m=1}^{K} H_{im}$. Therefore, the normalized version of transfer function element H_{ij} becomes $H_{ij} / \sum_{m=1}^{K} H_{im}$. Because H_{ij} is usually a complex number, this ratio is further simplified to the squared magnitude, generating the commonly used definition of the DTF γ_{ij}:

$$\boxed{\gamma_{ij} = \frac{|H_{ij}|^2}{\sum_{m=1}^{K} |H_{im}|^2}}$$ (7.7)

Thus, the DTF can be determined from transfer matrix H, which can be determined using the matrix of AR coefficients A (Equation (7.5c)). A recent study by Wilke et al. (2009) describes how DTF can be combined with adaptive parameter estimation; this adaptive version of DTF allows time-variant coefficients of the AR model to deal with nonstationarity of the EEG signal.

7.3.2 Implementation

After obtaining the above results, it is appropriate to start thinking about an algorithm to determine γ_{ij} from measured data (e.g., a multichannel EEG recording). One practical approach to find the transfer matrix H in a measured data set is to fit an AR model to the data and determine matrix A of the AR coefficients (Section 7.3.1.2). The inverse of A gives H (Equation (7.5c)). This is a parametric approach, because the basis is to fit parameters of an AR model to the data and the DTF is derived from there. Fitting a multichannel AR model to a data set is an art in itself that is beyond the scope of this chapter. If you want to play with fitting models, there are several Web sites with MATLAB scripts available; one example is the ARfit toolbox from http://www.gps.caltech.edu/~tapio/arfit/.

If you want to evaluate this arfit *code you can create your AR model and evaluate the performance of the estimator's output with the known coefficients in your model. The following script,* Pr7_1.m, *is an example of this procedure. Note that this works only if you download and install the* arfit *toolbox! Recall to include the* arfit *directory in the path by using "Set Path ..." in the "File" menu.*

```
% Pr7_1.m
% A test for identifying coefficients from a time series

% The program prints the coefficients we use (a and b)
% plus their estimates (vector A) obtained with the arfit function
% from the ARfit toolbox:http://www.gps.caltech.edu/~tapio/arfit/

clear;
close all;

% Set coefficients a and b
a=0.95;
b=-.55;
% Parameters & initial values for the time series
N=1000;
e=randn(1,N+3); % GWN input
x(1)=0;e(1)=0;
x(2)=0;e(2)=0;

% create the time series using autoregression
for i=3:N+3
    x(i)=a*x(i-1)+b*x(i-2)+e(i);
end;
% Remove the 1st 2nd zero-valued-points from x
x=x(3:N+3);
% Make e the same length
e=e(3:N+3);
% normalize x & e
x=x-mean(x);
x=x/std(x)^2;
e=e-mean(e);
e=e/std(e)^2;

% arfit toolbox
[w,A,C,SBC,FPE,th]=arfit(x',0,4);      % arfit function
```

```
% Show the results
('coefficients')
('in')
[a b]
('estimated')
A
```

When you run the above script you will find that the arfit routine finds reasonable estimates for the coefficients a and b (i.e., your estimates for a and b will be close to 0.95 and −0.55, respectively).

For the implementation we can also go the **nonparametric** route by assuming that the noise sources in our model are zero mean GWN signals. This gives us the expression in Equation (7.6c), which shows us that the unscaled power spectrum $S(f)$ is proportional to $[H(f)H(f)^*] = |H|^2$. Under this assumption, we may determine the power spectrum directly from the data without having to estimate parameters by using the proportionality between S and $|H|^2$ to estimate the DTF.

7.3.2.1 Examples

Let us simulate an example. First we create three signals S1, S2, and S3 with causal relationships S1 → S2 and S1 → S3.

In MATLAB we accomplish this by creating a delay between S1 and S2 (in our example 5 samples delay) and S1 and S3 (10 samples delay). The signal-to-noise ratio (SNR) of our signal in S1 is determined by the variable SNR, and the strength of the coupling of the signal in S1 to S2 and S3 is determined by K2 and K3, respectively (see top diagram in Fig. 7.2).

```
function [S1 S2 S3]=Simulated_Signal(SNR, K2, K3);
% Output signals S1, S2, S3
% the signal-to-noise for S1 is determined by SNR
% the coupling from S1 to S2 and S3 are determined by K2 and K3
% Delay between S2 and S1 is dly2 and between S3 and S1 is dly3
% Linear vs. nonlinear coupling is determined by flag NL
% parameters

sample_rate=400;      % sr in Hz
freq=30;              % f in Hz
tim=40;               % time in seconds
dly2=5;               % # points delay and coupling strength signal 1 ->2
dly3=10;              % # points delay and coupling strength signal 1 ->3

% Nonlinear Flag (coupling can be linear (0) or nonlinear (1))
NL=0;
```

```
% derived parms
f_Nyq=sample_rate/2;          % Nyquist
dt=1/sample_rate;             % time step
N=tim*sample_rate;            % no of points
t=0:dt:tim;                   % time axis
noise=randn(1,length(t));     % noise component
null=zeros(1,length(t));      % basline of zeros

% Source Signal
S1=sin(2*pi*freq*t).*SNR*std(noise);
% Create the derived signals S2 and S3
S2=null; S3=null;
for k=dly2+1:N
  if (NL==0);
    S2(k)=S1(k-dly2).*K2+randn;           % linear coupling
  else;
    S2(k)=(S1(k-dly2).^2).*K2+randn;      % nonlinear coupling
  end;
end;
for k=dly3+1:N
  if (NL==0);
    S3(k)=S1(k-dly3).*K3+randn;           % linear coupling
  else;
    S3(k)=(S1(k-dly3).^2).*K3+randn;      % nonlinear coupling
  end;
end;
% Noise added to S1
S1=S1+noise;
% Normalize all signals
S1=(S1-mean(S1))/std(S1);
S2=(S2-mean(S2))/std(S2);
S3=(S3-mean(S3))/std(S3);
```

In our example we run the above function by typing

```
[el1 el2 el3]=Simulated_Signal(3,0.1,0.001);
```

and the result is three traces simulating signals from electrodes 1−3 (el1, el2, and el3) with causal relationships; we then save the result in a file test.mat:

```
save test el1 el2 el3
```

Next, we use the file named test to investigate the causal relation between the signals with the script Pr7_2.m.

```
% Pr7_2.m
% Demo DTF based on signals generated with function Simulated_Signal,
% These signals are saved in File test.mat
% !! the function regres must be in the directory to detrend the data !!

% This program steps with 20 s windows (duration) through
% each of three 40 s signals (el1, el2, el3).
% These 20 second windows move ahead with steps of 5 s (increment).
% So there is 15 s overlap between the 20 s analysis windows)
% Within each window of 20 seconds, the average (cross)spectra
% are computed from fft-analysis epochs of 128 points (step).

% NOTE: THIS PROGRAM IS NOT OPTIMIZED.

clear;
load test                       % load the data with 40 s input traces el1 - el3

% Parameters
cmd0=['N=length(el1)'];         % Determine the length of the signal
eval([cmd0 ';'])
sample_rate=400;                % 400 Hz sample rate
duration=20;                    % duration of the total analysis window in
                                % seconds
step=128;                       % # of points in the FFT analysis window
increment=5;                    % steps of the analysis window in seconds
dt=1/sample_rate;               % sample interval
fNyq=sample_rate/2;             % Nyquist frequency
df=1/(step*dt);                 % Frequency step for the FFT
f=0:df:fNyq;                    % Frequency axis for the FFT

% Plot the three signals el1 - el3 in the top panels
figure
subplot(4,3,1);
plot(el1);hold;axis([0 N min(el1) max(el1)]);
t=['el1'];title(t);
axis('off');
subplot(4,3,2);
plot(el2);hold;axis([0 N min(el2) max(el2)]);
t=['el2'];title(t);
axis('off');
subplot(4,3,3);
plot(el3);hold;axis([0 N min(el3) max(el3)]);
t=['el3'];title(t);
axis('off');
```

```
% MAIN LOOP: STEPPING THROUGH THE DATA
% AND COMPUTING THE (CROSS)SPECTRA
count=0;
for w=1:increment*sample_rate:N-duration*sample_rate
   % Move data window into x, y, z
   x=el1(w:w+duration*sample_rate-1);
   y=el2(w:w+duration*sample_rate-1);
   z=el3(w:w+duration*sample_rate-1);

   % Initialize the Cross-Spectral arrays for averaging
   Sxx=zeros(1,step);
   Syy=Sxx;
   Szz=Sxx;
   Sxy=Sxx;
   Sxz=Sxx;
   Syz=Sxx;

   % SECOND LOOP TO COMPUTE AVERAGE (CROSS)SPECTRA
   xtemp=0:step-1;
   for i=1:step:sample_rate*duration-step;
      % pre-processing x
      [m,b,r]=regres(xtemp,x(i:i+step-1));      % Use regression to compute trend
      trend=m*xtemp+b;
      x(i:i+step-1)=x(i:i+step-1)-trend;                      % detrend
      x(i:i+step-1)=x(i:i+step-1)-mean(x(i:i+step-1));    % demean
      fx=fft(x(i:i+step-1).*hann(step)');                      % windowed fft
      % pre-processing y
      [m,b,r]=regres(xtemp,y(i:i+step-1));
      trend=m*xtemp+b;
      y(i:i+step-1)=y(i:i+step-1)-trend;
      y(i:i+step-1)=y(i:i+step-1)-mean(y(i:i+step-1));
      fy=fft(y(i:i+step-1).*hann(step)');
      % pre-processing z
      [m,b,r]=regres(xtemp,z(i:i+step-1));
      trend=m*xtemp+b;
      z(i:i+step-1)=z(i:i+step-1)-trend;
      z(i:i+step-1)=z(i:i+step-1)-mean(z(i:i+step-1));
      fz=fft(z(i:i+step-1).*hann(step)');
      % compute all 9 spectra which are proportinal with |H|^2, Eq (7.6c)
      Sxx=Sxx+fx.*conj(fx);
      Syy=Syy+fy.*conj(fy);
      Szz=Szz+fz.*conj(fz);
      Sxy=Sxy+fx.*conj(fy);
```

```
    Sxz=Sxz+fx.*conj(fz);
    Syz=Syz+fy.*conj(fz);
    Syx=conj(Sxy);
    Szx=conj(Sxz);
    Szy=conj(Syz);
end;

% Compute the power
S11=abs(Sxx).^2;
S12=abs(Sxy).^2;
S13=abs(Sxz).^2;
S21=abs(Syx).^2;
S22=abs(Syy).^2;
S23=abs(Syz).^2;
S31=abs(Szx).^2;
S32=abs(Szy).^2;
S33=abs(Szz).^2;

% Normalize
NS11=S11./max(S11);        % on diagonal the normalized power spectrum
NS12=S12./(S11+S12+S13); % Eq (7.7)
NS13=S13./(S11+S12+S13); % Eq (7.7)
NS21=S21./(S21+S22+S23); % Eq (7.7)
NS22=S22./max(S22);        % on diagonal the normalized power spectrum
NS23=S23./(S21+S22+S23); % Eq (7.7)
NS31=S31./(S31+S32+S33); % Eq (7.7)
NS32=S32./(S31+S32+S33); % Eq (7.7)
NS33=S33./max(S33);        % on diagonal the normalized power spectrum

count=count+1;

% Plot the results in the corresponding panels and
% superimpose the results for different epochs

% Titles for the panels
ttle1=[' ' num2str(count) ' ' 'Spectrum el1'];
ttle2=' el2 -> el1';
ttle3=' el3 -> el1';
ttle4=' el1 -> el2';
ttle5=' Spectrum el2';
ttle6=' el3 -> el2';
ttle7=' el1 -> el3';
ttle8=' el2 -> el3';
ttle9=' Spectrum el3';
```

```
% Draw a red horizontal line for each 20 s analysis window
Y=[0 0];
X=[w w+duration*sample_rate];
XP=[w+1-increment*sample_rate w];
subplot(4,3,1);plot(X,Y,'r');if (count > 1);plot(XP,Y);end;
subplot(4,3,2);plot(X,Y,'r');if (count > 1);plot(XP,Y);end;
subplot(4,3,3);plot(X,Y,'r');if (count > 1);plot(XP,Y);end;
% Plot the (cross)spectral information in the lower 3 × 3 panels
subplot(4,3,4);hold on; plot(f(1:step/4),NS11(1:step/4),'k');axis([0 60 0 1]);
title(ttle1);
subplot(4,3,5);hold on; plot(f(1:step/4),NS12(1:step/4),'k');axis([0 60 0 1]);
title(ttle2);
subplot(4,3,6);hold on; plot(f(1:step/4),NS13(1:step/4),'k');axis([0 60 0 1]);
title(ttle3);
subplot(4,3,7);hold on; plot(f(1:step/4),NS21(1:step/4),'k');axis([0 60 0 1]);
title(ttle4);
subplot(4,3,8);hold on; plot(f(1:step/4),NS22(1:step/4),'k');axis([0 60 0 1]);
title(ttle5);
subplot(4,3,9);hold on; plot(f(1:step/4),NS23(1:step/4),'k');axis([0 60 0 1]);
title(ttle6);
subplot(4,3,10);hold on; plot(f(1:step/4),NS31(1:step/4),'k');axis([0 60 0 1]);
title(ttle7);
subplot(4,3,11);hold on; plot(f(1:step/4),NS32(1:step/4),'k');axis([0 60 0 1]);
title(ttle8);
subplot(4,3,12);hold on; plot(f(1:step/4),NS33(1:step/4),'k');axis([0 60 0 1]);
title(ttle9);

% Force the script to draw the plots and pause for 1 second
drawnow;
pause(1);

% END MAIN LOOP
end;
```

The final result of running the above scripts is a plot as depicted in Fig. 7.2. The top row of plots in Fig. 7.2 shows the three signals in the time domain and the bottom part (3 × 3 matrix of panels) shows the spectral plus "causal" information. The diagonal panels are the spectra of each of the three channels and the off-diagonal plots show the DTF. It can be seen that, as expected, the first columns shows a consistent energy peak around 30 Hz (the frequency of the test signal in el1) in the DTF, reflecting the causal relationships between el1 and the other two electrodes (el2 and el3). The spectral panels show superimposed traces because spectra and DTFs were determined a number of times for five subsequent epochs of the signals.

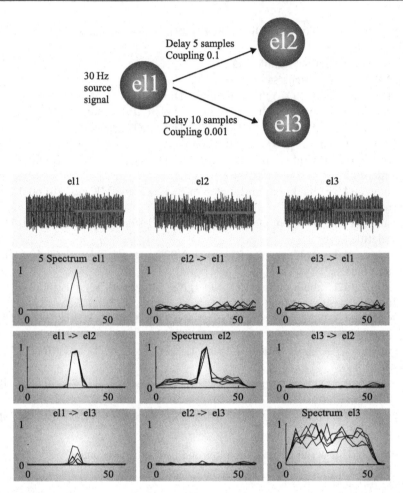

Figure 7.2 Test of the nonparametric DTF algorithm on simulated data. The top diagram shows how three signals el1, el2, and el3 relate to each other. The source el1 contains a 30 Hz sinusoidal signal that is coupled with delays to the other two electrode signals el2 and el3. Noise is added to all three channels. The top panels el1, el2, and el3 show the signals plus their noise components in the time domain and the bottom 3 × 3 panels are the result of the DTF analysis. These panels show multiple traces, the results from analyzing five overlapping epochs superimposed. Each epoch is 20 s and the overlap between subsequent epochs is 15 s (the red horizontal lines in the three upper panels represent the 20 s analysis windows used for the last of the five epochs). The diagonal panels in the 3 × 3 arrangement show the power spectra scaled between 0 and 1 of each electrode and the off-diagonal panels show the DTF according to Equation (7.7). It is clear that the first column contains energy around 30 Hz, confirming that el1 is a source for el2 and el3. The graphs were prepared with MATLAB scripts Simulated_Signal.m and Pr7_2.m; if you repeat this procedure your results may slightly differ (due to the effects of the added noise components).

Recording of brain electrical activity from the scalp (EEG) or the cortex (ECoG) is the clinical basis for the evaluation of patients with epilepsy. These recordings can capture interictal spikes and the epileptic seizures and they can be used to determine the temporal and spatial characteristics of these activity patterns. For surgical candidates, a precise localization of the region where seizures originate is highly significant because it determines the target for surgical resection. Therefore, it is clinical practice to monitor surgical candidates for several days. In such clinical recordings, even if we assume that the electrodes sufficiently cover the brain areas of interest, the determination of the origin of the ictal activity can be far from simple. First, the epileptologist must detect all seizures occurring in a large data set; second, within each seizure the origin of the epileptiform discharges must be determined. To determine the epileptic focus, the epileptologist will use multiple data sets reflecting brain structure and function (EEG, MRI, PET, etc.). The DTF analysis is a natural fit into this set of clinical data because it provides an indicator where activity may originate.

7.4 Combination of Multichannel Methods

Finally we discuss how several of the multichannel techniques can be employed to investigate brain activity. Gómez-Herrero et al. (2008) developed and applied a novel methodology based on multivariate AR modeling and ICA to determine the temporal activation of the intracerebral EEG sources as well as their approximate locations. First these authors used PCA to remove noise components (similar to our example with Lena's image in Pr6_2.m) and ICA to identify the EEG sources (as in our example in Fig. 6.15). The direction of synaptic flow between these EEG sources is then estimated using DTF (as we did in the example of Fig. 7.2). The reliability of their approach is assessed with simulations and evaluated by analyzing the EEG-alpha rhythm. Their results suggest that the major generation mechanism underlying EEG-alpha oscillations consists of a strong bidirectional feedback between thalamus and posterior neocortex. Altogether, the study suggests that the combined application of PCA, ICA, and DTF is a promising noninvasive approach for studying directional coupling between neural populations.

References

Arfken, G.B., Weber, H.J., 2005. Mathematical Methods for Physicists, sixth ed. Academic Press, Elsevier, Burlington, MA.

Barbero, A., Franz, M., Van Drongelen, W., Dorronsoro, J.R., Scholköpf, B., Grosse-Wentrup, M., 2009. Implicit Wiener series analysis of epileptic seizure recordings. Conf. Proc. IEEE Eng. Med. Biol. Soc. 1, 5304–5307.

Bell, A.J., Sejnowski, T.J., 1995. An information-maximization approach to blind separation and blind deconvolution. Neural Comput. 7, 1129–1159.

Boas, M.L., 1966. Mathematical Methods in the Physical Sciences, second ed. John Wiley & Sons.

Cover, T.M., Thomas, J.A., 1991. Elements of Information Theory, John Wiley & Sons, New York.

De Schutter, E., Bower, J.M., 1994a. An active membrane model of the cerebellar Purkinje cell. I. Simulation of current clamps in slice. J. Neurophysiol. 71, 375–400.

De Schutter, E., Bower, J.M., 1994b. An active membrane model of the cerebellar Purkinje cell. II. Simulation of synaptic responses. J. Neurophysiol. 71, 401–419.

Fitzhugh, R.A., 1961. Impulses and physiological states in theoretical models of nerve membrane. Biophys. J. 1, 445–466.

Franz, M.O., Schölkopf, B., 2006. A unifying view of Wiener and Volterra theory and polynomial kernel regression. Neural Comput. 18, 3097–3118.

Gómez-Herrero, G., Atienza, M., Egiazarian, K., Cantero, J.L., 2008. Measuring directional coupling between EEG sources. Neuroimage 43, 497–508.

Granger, C.W.J., 1969. Investigating causal relations by econometric models and cross-spectral methods. Econometrica 37 (3), 424–438.

Hodgkin, A.L., Huxley, A.F., 1952. A quantitative description of membrane current and its application to conduction and excitation in the nerve. J. Physiol. 117, 500–544.

Izhikevich, E.M., 2007. Dynamical Systems in Neuroscience: The Geometry of Excitability and Bursting, MIT Press, Cambridge, MA.

Jordan, D.W., Smith, P., 1997. Mathematical Techniques. Oxford University Press, Oxford.

Kamiński, M., Ding, M., Truccolo, W.A., Bressler, S.L., 2001. Evaluating causal relations in neural systems: Granger causality, directed transfer function and statistical assessment of significance. Biol. Cybern. 85, 145.

Kamiński, M.J., Blinowska, K.J., 1991. A new method of the description of the information flow in the brain structures. Biol. Cybern. 65, 203.

Koch, C., 1999. Biophysics of Computation: Information Processing in Single Neurons, Oxford University Press, New York.

Krausz, H.I., 1975. Identification of nonlinear systems using random impulse train inputs. Biol. Cybern. 19, 217–230.

Lay, D.C., 1997. Linear Algebra and its Applications, Addison-Wesley, New York.

Lee, Y.W., Schetzen, M., 1965. Measurement of the kernels of a nonlinear system by cross-correlation. Int. J. Contr. 2, 237–254.

Lomb, N.R., 1976. Least-squares frequency analysis of unequally spaced data. Astrophys. Space Sci. 39, 447—462.

Lopes da Silva, F.H., Hoeks, A., Smits, H., Zetterberg, L.H., 1974. Model of Brain Rhythmic Activity: The Alpha-Rhythm of the Thalamus. Kybernetik 15, 27—37.

Marmarelis, P.Z., Marmarelis, V.Z., 1978. Analysis of Physiological Systems: The White Noise Approach, Plenum Press, New York.

Marmarelis, V.Z., 2004. Nonlinear Dynamic Modeling of Physiological Systems, IEEE Press, John Wiley & Sons Inc., Hoboken, NJ.

Martell, A., Lee, H., Ramirez, J.M., Van Drongelen, W., 2008. Phase and frequency synchronization analysis of NMDA-induced network oscillation. P142, CNS 2008 Annual Meeting. http://www.biomedcentral.com/content/pdf/1471-2202-9-s1-p142.pdf.

Pikovsky, A., Rosenblum, M., Kurths, J., 2001. Synchronization: A Universal Concept in Nonlinear Sciences, Cambridge University Press, Cambridge, UK.

Press, W.H., Teukolsky, S.A., Vetterling, W.T., Flannery, B.P., 2007. Numerical Recipes in C, third ed. Cambridge University Press, Cambridge, MA.

Recio-Spinoso, A., Temchin, A.N., van Dijk, P., Fan, Y-H., Rugero, M.A., 2005. Wiener-kernel analysis of responses to noise of chinchilla auditory-nerve fibers. J. Neurophysiol. 93, 3615—3634.

Scargle, J.D., 1982. Studies in astronomical time series analysis. II. Statistical aspects of spectral analysis of unevenly spaced data. Astrophys. J. 263, 835—853.

Schetzen, M., 2006. The Volterra & Wiener Theories of Nonlinear Systems, second reprint ed. Krieger Publishing Company, Malabar, FL.

Shannon, C.E., Weaver, W., 1949. The Mathematical Theory of Communication, University of Illinois Press, Urbana, IL.

Stone, J.V., 2004. Independent Component Analysis: A Tutorial Introduction, MIT Press, Cambridge, MA.

Traub, R.D., Contreras, D., Cunningham, M.O., Murray, H., LeBeau, F.E.N., Roopun, A., et al., 2005. Single-column thalamocortical network model exhibiting gamma oscillations, sleep spindles, and epileptogenic bursts. J. Neurophysiol. 93, 2194—2232.

Van Drongelen, W., 2007. Signal Processing for Neuroscientists: An Introduction to the Analysis of Physiological Signals, Academic Press, Elsevier, Amsterdam.

Van Drongelen, W., Koch, H., Elsen, F.P., Lee, H.C., Mrejeru, A., Doren, E., et al., 2006. The role of persistent sodium current in bursting activity of mouse neocortical networks in vitro. J. Neurophysiol. 96, 2564—2577.

Van Drongelen, W., Williams, A.L., Lasky, R.E., 2009. Spectral analysis of time series of events: effect of respiration on heart rate in neonates. Physiol. Meas. 30, 43—61.

Westwick, D.T., Kearney, R.E., 2003. Identification of Nonlinear Physiological Systems, IEEE Press, John Wiley & Sons Inc., Hoboken, NJ.

Wilke, C., Van Drongelen, W., Kohrman, M., He, B., 2009. Identification of epileptogenic foci from causal analysis of ECoG interictal spike activity. Clin. Neurophysiol. 120, 1449—1456.

Zinn-Justin, J., 2002. Quantum Field Theory and Critical Phenomena, Oxford University Press, New York.

Printed in the United States
By Bookmasters